T0316671

The End of Infinity

The End of Infinity

Where Mathematics and Philosophy Meet

Anthony C. Patton

Algora Publishing
New York

Library of Congress Cataloging-in-Publication Data —

Names: Patton, Anthony C., 1969- author.
Title: The end of infinity: where mathematics and philosophy meet
/ Anthony C. Patton.
Description: New York: Algora Publishing, 2018. | Includes bibliographical
references and index.
Identifiers: LCCN 2018015141 (print) | LCCN 2018016394 (ebook) | ISBN
9781628943412 (pdf) | ISBN 9781628943399 (soft cover: alk. paper) | ISBN
9781628943405 (hard cover: alk. paper)
Subjects: LCSH: Infinite.
Classification: LCC BD411 (ebook) | LCC BD411 .P38 2018 (print) | DDC
111/.6—dc23
LC record available at https://lccn.loc.gov/2018015141

Printed in the United States

Table of Contents

Cast of Characters

1. Natural Numbers: The set of positive numbers (1, 2, 3, ...).

2. Integers: The set of positive and negative numbers (... , -3, -2, -1, 0, 1, 2, 3, ...).

3. Rational Numbers: The set of positive and negative numbers and fractions (1, 2, -3, 1/2, 1/4, -1/8, ...). Every rational number can be expressed as the ratio of two integers. Rational numbers can also be expressed as decimals (0.5, 0.25, -0.125, ...).

4. Real Numbers: The set of rational and irrational numbers that is often thought of as the set of numbers for a continuum. Irrational numbers cannot be expressed as the ratio of two integers, such as $\sqrt{2}$ or π.

5. Dense: A property of rational and real numbers, such that there always exists a number between any two rational or real numbers. For example, 1/2 is between 0 and 1; 3/4 is between 1/2 and 1; 7/8 is between 3/4 and 1; and so on, ad infinitum. As a result, rational or real numbers are not sequential like the natural numbers or integers. (There is a creative way to resolve the density problem to list the numbers sequentially.)

6. $\aleph_0$: According to Cantor, the smallest infinite set that is equal to the set of natural numbers (1, 2, 3, ...) and all other infinite sets that are countable or denumerable with the natural numbers (expressed as $\aleph$ in this book).

7. Cardinality: The quantity of elements in a set. For example, the set (2, 4, 8) has a cardinality of 3 (even though the number 3 is not a member of the set).

8. Isomorph: A property of two sets that can be put into a one-to-one correspondence to demonstrate that they have the same cardinality. For example, we can pair up the even and odd numbers less than or equal to 10 to show that they have the same cardinality (5).

1→ 2

3→ 4

5→ 6

7→ 8

9→ 10

9. Infinite Set: According to Dedekind, any set that can be put in to a one-to-one correspondence or isomorph with a proper subset of itself. For example, every even number is a natural number but not every natural number is an even number, which makes the even numbers a proper subset of the natural numbers. For every natural number we can list an even number, ad infinitum.

10. Power Set: The set of all subsets of a set (all possible combinations). For example, for the set (2, 4, 6), the power set is [(0), (2), (4), (6), (2, 4), (2, 6), (4, 6), (2, 4, 6)]. If the power set contains n elements, the power set contains 2^n elements (2 x 2 x 2 x ... x 2).

11. Gödel's Incompleteness Theorem: A theorem that proved it is impossible to find a complete and consistent set of axioms for mathematics. The first theorem proved that a consistent system of axioms cannot prove all the truths of mathematics. The second theory proved that a system of axioms cannot demonstrate its own consistency. Gödel used this to prove that set theory is incomplete, which means that Cantor's foundation for his theory of infinity is incomplete.

Arithmetic is a computational science in its calculations, with true and perfect units, but it is of no avail in dealing with continuous quantity.

—Leonardo da Vinci

Introduction

As a mathematics major, I imbibed the idea of infinity with reckless abandon. I was proud of my elite membership and knew that students from other majors could only marvel at the alchemy we conjured behind closed doors. Calculus, for example, opened my mind to a powerful and abstract way of thinking with real world applications that the uninitiated could not fathom. To this day, incredulous heretics often ask me, "How often do you actually *use* Calculus?" My answer, then as now, is always the same: "Every day." Thinking mathematically has practical consequences.

My primary interest was applied mathematics, not pure mathematics, but I was intrigued by the idea of infinity, which is why my heart fluttered the first time I heard the paradoxical claim of a nineteenth-century mathematician named Georg Cantor who said infinity was real, that there was more than one type of infinity, and that not all infinities were the same size! As a member of the Cantor camp, this was just what I needed to hear to solidify my membership. Like Tertullian, I pledged to believe Cantor because it was absurd. I honestly had no idea what Cantor meant by this shocking claim, but I hoped I could impress others with this miraculous piece of wisdom.

I also majored in philosophy, which allowed me to consider infinity from another perspective that was equally important to the discussion. To my pleasant surprise, many of the greatest philosophers of the Western tradition were great mathematicians or reflected on mathematics and infinity. I was not surprised to hear that Descartes, the inventor of analytic geometry, or Leibnitz, the co-inventor of Calculus, were two of the most important philosophers of the modern age, or that the titans of ancient Greece, Plato and Aristotle, recognized the importance of mathematics and infinity more than 2,000 years ago. This reinforced my belief in the importance of mathematics and philosophy, and I soon learned that the line between mathematics and philosophy blurred at the most abstract levels, where the bells of theology chimed on the horizon.

After college my interests shifted to philosophy, and more than 25 years would pass before Cantor's theory of infinity would surface again. The infinity hangover had long worn off and I was willing to reconsider my old faith with a critical eye. After reading a variety of literature on Cantor's theory of infinity, I decided to do a deep dive into the mathematics and philosophy of infinity to get to the bottom of the issue one way or the other, because the consequences were important for mathematics and philosophy. My own intuition was telling me it was either one of the most absurd or one of the most important theories in intellectual history. Even Russell, a great proponent of Cantor's ideas, was honest about the fragile foundations of infinity.

"It cannot be said to be *certain* that there are in fact any infinite collections in the world. The assumption that there are is what we call the 'axiom of infinity.' Although various ways suggest themselves by which we might hope to prove this axiom, there is reason to fear that they are fallacious, and that there is no conclusive logical reason for believing them to be true. At the same time, there is certainly no logical reason *against* infinite collections, and we are therefore justified, in logic, in investigating the hypothesis that there are such collections."[1]

As we will see, there are logical reasons against infinite collections, especially in the context of mathematics. Before beginning my deep dive, I asked myself whether I was qualified to investigate the issue. After all, I studied mathematics and philosophy in college, but I was not a specialist in set theory, which was the framework Cantor used to prove his theory about infinity. Therefore, mindful that it would be an uphill battle to disprove Cantor's theory, I decided to look at the problem from two perspectives. First, given that Cantor made specific claims within framework of set theory (the inside approach), I set out to identify any problems or errors in these claims. For example, should we accept Cantor's conclusions about the arithmetic of in-

[1] Russell, Bertrand, *Introduction to Mathematical Philosophy*, 47.

finity or his calculations for the cardinality of the rational and real numbers? Is his isomorph or power set methodology valid? Second, given that Cantor claimed set theory provides a foundation for mathematics and that infinity is real (the outside approach), I set out to identify any problems or errors in these claims as well. For example, is it possible to show that set theory does not provide a foundation for mathematics or that the idea of infinity generates a contradiction? With these two paths before me, I set out with an open mind.

After conducting my deep dive, I can report with complete humility that the results were better than anticipated in terms of confirming my initial intuition that Cantor made errors. As it turns out, the argument about infinity boils down to two points: first, whether infinite sets exist (is the Axiom of Infinity valid?); and second, whether it is possible to "arithmetize" the continuum, that is, to transform the discrete into the continuous. All mathematical proofs about infinity are a waste of time until we resolve these two points. Of interest, however, I concluded that these are philosophical questions that cannot be solved with mathematics. Cantor concluded that both points are true, which allowed him to argue that the real numbers are not countable. In this book I will argue that both points are false and reveal to the Cantor camp the fragile assumptions at the foundation of their beliefs.

The following are some of the highlights that this book will address in more detail:

1. Set theory does not provide a foundation for mathematics because set theory is incomplete (according to Gödel's incompleteness theorem) and inconsistent (due to the contradictions proved in this book). Additionally, logic and mathematics are two distinct disciplines; one cannot be reduced to the other.

2. The idea of actual infinity has unavoidable metaphysical implications, regardless of our definitions. The ideas of actual infinity and dimensionless points are incompatible with a finite continuum and violate the law of contradiction because points cannot both have and lack dimension at the same time. It is not possible to evenly distribute an infinite set of points along a finite continuum.

3. The one-to-one correspondence or isomorph methodology (a set is infinite if it can be put into a one-to-one correspondence or isomorph with a

proper subset of itself) is invalid because the mere fact that we can begin the process of pairing up two sets of numbers does not mean they have the same cardinality. For example, we can begin the process of pairing up the natural numbers and the real numbers, but Cantor argues that they do not have the same cardinality.

4. Cantor claimed the cardinality of the rational numbers is $\aleph$ and the cardinality of the real numbers is $2^{\aleph}$, but there are valid methodologies to prove that the cardinality of both sets is $10^{\aleph}$. The fact that different yet valid methodologies generate different results proves that set theory is inconsistent. The fact that the cardinality of the real numbers is $10^{\aleph}$, not $2^{\aleph}$, means the continuum cannot be "arithmetized."

5. Cantor's claims that $\aleph \times \aleph = \aleph$ and $\aleph < 2^{\aleph}$ generates a contradiction. If $\aleph \times \aleph \times \aleph \times \ldots \times \aleph = \aleph$ and $2 \times 2 \times 2 \times \ldots 2 = 2^{\aleph}$, then $\aleph \times \aleph \times \aleph \times \ldots \times \aleph < 2 \times 2 \times 2 \times \ldots 2$, which means $\aleph < 2$. (The Cantor camp would deny that $2 \times 2 \times 2 \times \ldots 2 = 2^{\aleph}$, but their use of infinite binary decimals confirms it.) Given that this is not true, this means $\aleph \times \aleph = \aleph^2$, $\aleph \times \aleph \times \aleph = \aleph^3$, ... , and $\aleph \times \aleph \times \aleph \times \ldots \times \aleph = \aleph^{\aleph} > \aleph$. Given that multiplication is a form of addition, this means $\aleph + \aleph = 2\aleph$, $\aleph + \aleph + \aleph = 3\aleph$, ... , and $\aleph + \aleph + \aleph + \ldots + \aleph = \aleph\aleph = \aleph^2$. With this established, Cantor's distinction between the cardinality of the real numbers and countable infinite sets no longer holds. With this break down in the arithmetic of infinity, we can prove that the set of even numbers is not the same size as the natural numbers ($\aleph/2 < \aleph$).

One of the most important findings of my deep dive was the understanding that infinity is primarily a philosophical problem, not a mathematical problem. A theory of infinity is not required for mathematics and cannot be solved with mathematics. Just as we cannot use language to resolve the problems of language, we cannot use mathematics to resolve the problems of mathematics. Languages and numbering systems reflect the strengths and weaknesses of the human mind, and some problems and gaps are baked into the cake; they require philosophical analysis for clarity.

All academic disciplines rest on philosophical assumptions, to include mathematics, so we should devote some time to getting the philosophy right because our journey will take us to the wrong destination if our initial steps are without a solid foundation.

Prelude: Time

Many people are fascinated by Einstein's theory of relativity but could never provide a lecture on the topic, the author included. To state it crudely, the theory claims there is no universal time standard (a universal clock) that we can all set our watches to, and that time can "speed up" or "slow down" depending on different variables such as speed and gravity. There is no objective standard of time for people living in two different galaxies. We have all seen the example of a digital clock that comes back from space a tiny fraction of a second slower. Or, we can watch a movie like *Interstellar* in wonder as minutes lived on one planet in outer space equals years lived on earth. As fascinating as this might be, we should take a closer look at this paradox to untangle some of the key variables and plant some seeds for our analysis of infinity.

We experience time in two ways: our internal clock, which has objective and subjective elements, and an external clock, which operates independently of our internal clock, such as the rotation and orbit of the earth around the sun, of cyclical motion in the universe. (For the sake of brevity, I will not raise Kant's claim that time and space are *a priori* forms of the intellect.) Our heartbeat is objective because we can count it, but it varies from moment to moment. People who have significant emotional events (fear, love, near death experience, etc.) are usually surprised after the fact by how much time has passed. The most common experience is of time slowing down, such that what seems like a minute really lasts only seconds. This is especially common when people are in danger. We are told that nature "slows down" time for us to promote our survival, like Neo in *The Matrix* leaning back to avoid a barrage of bullets. On the other hand, for people who are the opposite of caught up in the moment (bored), time drags and cannot go fast enough. People around the world glance at clocks throughout the day, eager for the workday to end so they can finally do what they really enjoy, only to have time speed up and rush them forward to the next workday.

For immensely practical yet nonetheless arbitrary reasons, we have selected a combination of the earth's rotation and orbit around the sun as the standards for our measure of time. The rotation and orbit of Jupiter is also real, but would be of limited use to us on earth. Man is the measure of all things. The earth rotates 365 times ("days") to complete one orbit around the sun ("year"). Also for practical yet nonetheless arbitrary reasons, we have selected the numbers 24 (hours) and 60 (minutes and seconds) to further divide the fractions of a day. These two numbers are practical because they are easily divisible by many numbers, which makes the mathematics of fractions easier (60 is divisible by 1, 2, 3, 4, 5, 6, 10, 12, 15, 20, 30, and 60).

One day has 24 hours; one hour has 60 minutes; and one minute has 60 seconds. Thus, one year consists of 31,536,000 seconds. These baseline units of seconds, minutes, hours, days, and years have the right level of granularity to put most human activity into a meaningful context—not too long, not too short, just right. These units also allow us to measure time in big numbers and small numbers, but the ability to do this does not tell us anything about the universe. For example, there might be no natural limit to how small we can divide a numerical fraction, but there might be a natural limit to how small we can divide the physical world—a smallest physical distance.

This brings us to a profound question: if one second equals 1/31,536,000 of the earth's orbit around the sun, how can time ever be relative? When the earth revolves around the sun next year, each second will remain 1/31,536,000 of a year, by definition, just as 1/4 of a pie this year will be 1/4 of a pie next near. Time might be relative, but mathematics cannot be. If a team of astronauts travels to outer space on 1 January and returns to earth on 31 December, we will be irrefutably correct in saying they traveled for one year, regardless of their experience of time. Their travel to outer space will not change the rate at which the earth rotates or orbits around the sun.

However, what if we were to analyze the DNA telomeres of the astronauts upon their return and their identical twin siblings and find conclusive evidence that the ageing process for the astronauts was different than for their identical twin siblings on earth during the previous year? In other words, what if time as an ageing process "slowed down" or "sped up" for the astronauts? The claim makes intuitive sense (ageing in theory could speed up or slow down), but it forces us to revise how we define and understand time. Our unit of time (second) cannot be relative because it is a human abstraction based on the rotation and orbit of earth. We might have to make minor tweaks to the definition of a second if there are changes to the rotation or orbit of the earth around the sun, but there is no evidence to suggest that this will happen any time soon.

The obvious truth is that something other than the rotation and orbit of the earth around the sun shapes how we age. We can measure our age in earth years, but this objective standard does not correspond precisely with the subjective ageing process, not to mention our subjective experience of time, and time can never be thought of as relative if we use earth years as a unit of time. Some people age faster or slower than other people, but one earth year is one earth year today and will be next year. We could use earth years to track time in a different galaxy, but it would probably be of limited value. A unit of time works best when it "fits" with whatever is measuring time.

Socials scientists are prone to swoon when they hear about a coffee farmer in a quaint Andean mountain village who measures time by the number of cigarettes he smokes. When asked how long it takes to pick a row of coffee beans, he might say, "one cigarette." Never mind how long the average row is, how long it takes on average to smoke a cigarette, or how long on average he waits between puffs, careful observation might confirm that the farmer indeed smokes one cigarette for each row of picking coffee beans. We can only shake our heads and smile enviously, knowing that this fortunate man is not a slave to objective time like the rest of us. In fact, he can probably tell us how many cigarettes he will smoke to walk to the next village.

When I was in college we had a tradition of drinking two beers after a long day of classes and homework. We did not take a 30-minute break or a one-hour break; we took a two-beer break. We never checked the time, aside from glancing at how much beer we had in the bottle, and we had no illusions that drinking two beers correlated with a fixed duration of time, but our drinking rate was a reflection of relevant variables. We would discuss some things during the first beer, and the ritualistic transition to the second beer would take on a meaning of its own as we began the process of winding down the break. If we had more homework to do and it was late, we might hurry the process, but if we were done for the day and had time to relax, we might stretch it out to enjoy each other's company. The only hard and fast rule was two beers.

If we consider a scenario in which the "burn rate" of smoking, drinking, or ageing is a better way to understand time in a human context that is both objective in some ways and allows for the phenomenon of relativity to make sense, we could look for a standardized "burn rate." For example, in the case of an atomic clock, one second is defined as the duration of 9,192,631,770 periods of the radiation corresponding to the transition between the two hyperfine levels of the ground state of the caesium-133 atom, according to the United States Naval Observatory. I have no idea what that means, but although it defines one second in a different way, and is therefore nonetheless arbitrary, it opens up the possibility of the theory of relativity because this atomic clock burn rate might vary in different conditions.

The same goes for a human. Like the movie *Interstellar*, the rate at which we age (our burn rate) could speed up or slow down depending on things like speed and gravity; however, this does not mean time is relative any more than the fact that a log burns faster in a blast furnace than it does in a crackling winter campfire means that time is relative. The only way to establish an objective unit of time, and therefore give rise to the paradox of the theory of relativity, is to select an arbitrary unit of measure, with the understanding that we should not take this arbitrary unit as a foundation for time or use it to create paradoxes that only confuse us. That is, the apparent paradox of the theory of relativity boils down to the error of focusing on the wrong variables and using arbitrary units of measure to understand natural phenomenon.

The same idea could apply to a mechanical object like a clock. A clock does not have an organic burn rate like a cigarette or DNA telomeres, but the rate at which the springs and cogs move the hands on the clock is also a function of variables like speed and gravity. The time difference between a clock on earth and a clock in space might be trivial but it is real nonetheless. The paradox arises from selecting an arbitrary unit of time, like an hour, and then building an organic or mechanical clock to measure or track that unit of time, forgetting that the burn rate or mechanical process of the clock is a function of the environment where time is being measured. As we will see, and this is one of the key points of the book, the theory of infinity in many ways makes the arbitrary units of mathematics (numbers) more real than the world that provides the foundation for them.

Another way to measure time would be to identify specific biological events, such as puberty or brain development milestones, to track the life of a human organism, in which case the basis for calling someone an adult would reflect these objective biological indicators, not the arbitrary milestone of 18 orbits around the sun. This would probably cause more problems than it would solve for society, but it would better correspond with our common observation that not all people age the same way.

I hope it is clear how this analysis of time relates to the study of mathematics and infinity. The burn rate is a more natural way to understand time and allows us to better understand the theory of relativity, but absent the existence of cyclical motion to help us establish an arbitrary unit of time, like a second, there is no way to establish an objective unit of burn rate time, aside from saying one thing burns faster or slower than something else. However, for time and mathematics, even though we benefit from and often require arbitrary numbering systems to make calculations, we should not allow these arbitrary numbering systems to shape the way we think about the world. There is no limit to how high we can count, but this does not mean there is no limit to how many years the universe has existed and it does not mean actual infinity is real. Likewise, there is also no limit to how much we can divide a fraction, but this does not mean there is no limit to how much the world or a burn rate can be divided or that actual infinity is real. Our system of mathematics helps us quantify the world and helps us make predictions, but we should not put the cart before the horse by thinking the numbers are more fundamental than the world that gives rise to them.

With this prelude of time out of the way, we can now shift our analysis to mathematics and infinity.

Chapter 1: History

The history of mathematics is the story of humans struggling to understand and define mathematics, in particular, how to account for the often astonishing correlation between mathematics and the world and to understand how mathematics fits into the hierarchy of knowledge. The history of mathematics is filled with examples of new discoveries within the ivory tower of reflection that only later were found to be useful in solving practical problems in science. On the one hand, some argue mathematics works because it is derived from the world, the same way that language works because it is derived from the world. For example, the reason we can use numbers to count objects is because we use the plurality of objects in the world as the foundation for our numbering system. Or, the reason we can use language to talk about the objects in the world is because we use the plurality of objects in the world as the foundation for our system of language. On the other hand, especially for people who have studied advanced mathematics, there is an eerie sense that mathematics shapes the way the world operates and that it is our challenge to "discover" mathematical truths, which raises metaphysical questions for philosophy and suggests that we can think about mathematics from top-down or bottom-up perspectives.

A detailed history of mathematics could fill volumes, but for the sake of brevity I will focus on some key ideas and key thinkers to provide a foundation for our analysis of infinity. For example, the history of numbers and the trial-and-error process of making the transition from notches on a bone to base-10, place-value numbers is interesting, spanning from prehistory to

Babylonia, Egypt, and Greece, but in this book we will have to be content to summarize this intellectual leap in a few sentences.

The Pre-Socratics

Our story begins in sixth century B.C. in Ionia, a region of the ancient Greek world on the west coast of modern day Turkey that gave birth to Homer and Western philosophy. Mathematics was popular in Egypt and Babylonia, but the Greeks were the first to fully abstract mathematics in a systematic way to make it a general science that transcended particular situations. The earliest Greek philosophers, who were known as the pre-Socratics because they preceded Socrates, were especially impressed with the fact of change—birth, growth, decay, death, etc.—which gave rise to the problem of the One and the Many. (If you want to impress people at a dinner party, you can tell them you are studying the pre-Socratics.) We see a plurality of objects in the world that is defined by change and becoming, which suggests there must an underlying eternal being or permanence that holds it all together. The great insight of the pre-Socratics was that despite all the change we see, there must be something permanent; change is from one thing into something else. Something primary must persist despite all the change.

The first pre-Socratics thought of the world as entirely material, with no distinction made between spirit and matter or even between philosophy and science. When Thales postulated that all was water, he literally believed that everything, including people, could be reduced to the element of water, the primal substance underlying the world of change. They focused on the material objects of knowledge, the world out there, not on the subjects of experience (themselves) or whether knowledge was even possible. That is, they were dogmatic about their ability to discern truth about the world without asking whether or how knowledge was possible. Given this focus on trying to explain the material world, they are properly called cosmologists.

However, the pre-Socratics also arrived at conclusions by speculative reason and abstract ideas regarding a law-governed universe. Our raw perceptions of the world do not suggest to us an underlying unity. Such an insight requires an intellectual leap beyond perception. They did not solve the problem of the One and the Many, but they got us moving in the right direction in ways that are still relevant today, which is why they are rightly considered the first philosophers of the Western tradition.

With the pre-Socratics, the Western tradition begins making the transition from myth to science. Rather than attempt to explain events in nature by appealing to the gods or myths, the pre-Socratics started a journey down the road of science and philosophy, which also influenced the development of mathematics. In fact, we cannot understand mathematics unless we understand how it fits under the umbrella of philosophy. Thus, the pre-Socratics are important for this book for two reasons: first, some of the pre-Socratics were talented mathematicians who made notable contributions to mathematics; and second, the ideas of the pre-Socratics were influential in shaping the philosophies of Plato and Aristotle, who played a critical role in shaping our understanding of mathematics.

The Milesian School

The 28th of May, 585 B.C., is often cited as the day that Western philosophy began, because this is when Thales of Miletus, one of the Seven Wise Men of ancient Greece, predicted an eclipse. If true, this would suggest that Thales was versed in mathematics and astronomy. Other anecdotes of Thales include that he made an almanac, that he established a means of steering ships by the Little Bear constellation, and that he cornered the olive market by predicting a scarcity (which would make him the first known financial speculator). In the context of philosophy, he was famous for saying water was the ultimate stuff of the world. Exactly what he meant by this is beyond the scope of this book, but the important point is that he attempted to identify a foundation for the changing objects of the world.

Anaximander, the second of the Melesians, also thought the world must be made of ultimate stuff. However, rather than selecting one of the traditional four elements, such as water, earth, air, or fire, he made the abstract conclusion that the ultimate stuff of the world must be more ultimate than the four elements because something must give rise to the four elements, which was a profound insight, then and now. The Greeks thought of the four elements as conflicting opposites (hot–cold, wet–dry, etc.). Therefore, we need an explanation for what gave rise to these four elements and these conflicts, because if the four elements are in conflict, one of them cannot be more primary than the others. Something must give rise to the conflicting opposites.

Anaximander concluded that the ultimate stuff of the universe was unlimited, infinite, and indeterminate, what he called the apeiron, or απειρον in Greek. Obviously, any mention of infinity this early in human thought, however vague, is of interest to the history of mathematics. The apeiron, the substance without limits, was eternal and encompassed all the worlds. The Greeks did not believe in the Big Bang or in the creation of the universe *ex nihilo*, so the apeiron was considered eternal and unchanging. The apeiron was not aware of the world and did not interfere in the world, much like the God of deism. The objects we see in the world come from the apeiron and return to the apeiron, but this is an endlessly repeating process, an eternal recurrence. The popular nineteenth-century philosopher Nietzsche would make this idea a cornerstone of his philosophy.

Anaximander's apeiron was important because he looked for an ultimate stuff that transcended our senses and provided a foundation for the material world, but it is a long way from our traditional understanding of God. In fact, thinking about infinity not in terms of quantity but in terms of "unlimited" or "indeterminate" was surprisingly insightful and will be raised again later when we consider what makes the universe and mathematics possible. Just as a detective begins to formulate a theory about a crime based on the available clues, Anaximander reached the conclusion that we cannot talk about the material world without talking about what gave rise the material world, which is something we cannot perceive directly with our senses. Anaximander's speculation was shaped by Greek mythological and religious traditions, in particular, the idea of the creation of the world with a rational principle imposing order on a primeval chaos, but his ideas planted an important seed for Western philosophy.

The Pythagorean Society

We are now moving from Ionia to southern Italy, where the Pythagorean Society played a critically important role for the history of mathematics and philosophy. Pythagoras, like the Milesians, was from Ionia. He founded a religious community at Crotona in southern Italy in the second half of the sixth century B.C. The Pythagorean Society was known for being ascetic and religious, with rituals of purification and the practice of silence, similar to Orphism, but with a scientific spirit. Of interest, the Pythagoreans believed in metempsychosis, the transmigration or reincarnation of souls after death, which means they believed in the immortality of the soul and its separateness from the material body. This was a notable difference from the Milesians. As part of this belief, the Pythagoreans were prohibited from eating meat, partly because they never knew who they might eat for their next meal. Legend has it that Pythagoras recognized his dead friend's voice in the bark of a dog.

Aristotle said the Pythagoreans were devoted to mathematics, were the first to advance the study of mathematics, and proposed that mathematics was the principle of all things. To understand what this means, consider music. As the Pythagoreans discovered, musical notes can be represented as mathematical ratios. If we make a string half as long and pluck it, we raise the note an octave. If we double the length of a string, we lower the note an octave, and so on. The Pythagoreans believed this principle applied to all things in the world. They believed the objects of the world are made of numbers, and this is why the mathematical ratios work. The numbers of the Pythagoreans were limited to natural numbers (1, 2, 3, ...) and ratios of natural numbers (1/3, 5/8, 12/5, ...), what we today call the rational numbers.

As with Anaximander, the apeiron (the indefinite, infinite, and indeterminate) was an important idea for the Pythagoreans. However, they took this concept to the next level by combining it with the idea of a limit, peras (περας in Greek), to give form to the apeiron. This peras provided the tempering that is required for the harmony we see in the universe (order out of chaos), whether in music or in health, and these harmonic proportions can be expressed mathematically.

The Pythagoreans made important contributions to mathematics, especially in geometry, such as the Pythagorean Theorem. Examples of this formula were noted in Sumerian mathematics, but the Pythagoreans were the first to abstract the theorem as a general formula that could be used with all right triangles. Also of note, Books I, II, IV, VI, and probably III of Euclid's *Elements*, the most important mathematics book in history, were based on the work of the Pythagoreans.

Unfortunately, the Pythagorean theory of proportion soon faced serious problems. The discovery that some things in the world could not be expressed as a ratio of natural numbers caused a crisis in mathematics, which we will address later in this chapter. In short, however, the Pythagoreans made the mistake of attempting to give life to the human abstraction (numbers) that we use to understand the world in a quantitative way. If the Pythagoreans had focused on studying the deep patterns of the world and had understood that numbers are second order objects with no objective existence, they would have avoided this crisis and other problems.

The Eleatic School

The founder of the Eleatic School was Parmenides, who believed that being or permanence was real and that becoming or change was an illusion, which was in opposition to a philosopher named Heraclitus. Parmenides did not deny that we perceive change, but he rejected that it was real, similar to how the changes on a movie screen are an illusion, even if we perceive them as real. With this idea we have a distinction between truth and appearance, which was important in the history of philosophy. The Hindus have a similar view of the world with their idea of Maya. One of the disciples of Parmenides, Zeno, used this idea to construct arguments that are important for the history of mathematics and for thinking about infinity. Zeno's most important point was that everything is one because we generate paradoxes if we think of the world as a plurality.

Zeno addressed the paradoxes of plurality in different ways, but for the sake of simplicity we will focus on his bottom line as it relates to infinity. Many philosophers and mathematicians, then and now, have proposed that a finite continuum consists of infinitely many points. If this sounds unusual, think of how many possible numbers there are on a ruler between 0 and 1. Between any two points on a finite continuum there is another point, ad infinitum, which is another way of saying the points along a finite continuum are dense. For example, between 0 and 1 we have 1/2; between 1/2 and 1 we have 3/4; between 3/4 and 1 we have 7/8, and so on ad infinitum. Zeno's position assumes that a plurality of things in the world must be infinitely divisible and that we must divide them as far as possible to reach the point where it makes sense to count them, which seems to rule out the possibility of a finite plurality of objects in the world.

Therefore, if we assume that a finite continuum contains an infinite set of points, then the points either have dimension or lack dimension, according to the law of contradiction. If the points have dimension, then the infinite set of points will necessarily extend to infinity and will not fit on the finite continuum because any distance, no matter how small, multiplied by infinity equals infinity. On the other hand, if the points lack dimension, then the set of infinite points will collapse to zero and will not fit on the finite continuum because no matter how many times we add zero to zero, the final answer is still zero. (As we will see, some mathematicians have introduced the idea of infinitesimals to avoid this either or situation, but the logic remains.) In other words, there is no way to evenly distribute an infinite set of points along a finite continuum, which is an idea that will come front and center as we continue our analysis of infinity. The only ways to fill a finite continuum with points is to: 1. use a finite set of points with dimensions that touch end to end like billiard balls, or 2. (the same idea from the other perspective) use a finite set of points without dimension separated by spaces (ranges) with dimension. Neither option allows for the possibility of actual infinity without collapsing the finite continuum to zero.

Zeno also made arguments against the idea of motion that rely on similar ideas. For the first proof, Zeno supposes that we want to cross the street. If the Pythagoreans are correct, then we would have to cross infinitely many points in a finite time (notice the assumption about an infinite set of points). However, he concludes this is impossible. You therefore cannot cross the street, and therefore all motion is an illusion. For the second proof, Zeno supposes that Achilles and a tortoise are going to race and Achilles gives the tortoise a head start. When Achilles reaches the point where the tortoise had started the race, the tortoise will have advanced to a new point farther away. When Achilles reaches that point, the tortoise again will have advanced to another point, and so on. Thus, Achilles will always get closer to the tortoise but will never reach or surpass the tortoise. Thus, Zeno concludes that motion is an illusion. For the third proof, Zeno supposes that we shoot an arrow. According to the Pythagoreans, the arrow should occupy a given position in space. However, according to Zeno, to occupy a position in space is to be at rest. Therefore, the flying arrow is at rest, which is a contradiction.

Zeno believed that the apparent contradictions generated by these proofs supported the claim of Parmenides about the truth of being and permanence over the illusion of becoming and change. No matter how unusual these arguments might sound to the modern mind, they are loaded with important insights and assumptions that are relevant to the modern arguments about infinity. Of interest, Russell admired Zeno's arguments and viewed them as supporting his belief in actual infinity, but from my perspective Zeno's arguments discredit the idea of actual infinity in a way that no modern argument can overcome. To repeat, there is no way to evenly distribute an infinite set of points along a finite continuum. Often, the best solution to an argument is the one that eliminates paradox, such as recognizing that we live in a finite world with extension, in which case the idea of an infinite set of points evenly distributed along a finite continuum generates a logical contradiction.

The Crisis

Returning to the Pythagorean idea that the world is number, there is something deeply satisfying about the idea that the universe is based on number and harmony. It might be interesting to learn that we achieve octaves by halving or doubling the length of a string, but our hearts and minds are filled with wonder and awe if we learn that these same mathematical numbers and ratios provide the foundation for the world we live in. When we play a note on a piano, we are not merely plucking a string to achieve a desired vibration; we are tapping into a vibration of the universe itself. Perhaps this is why nature often reveals mathematical patterns? They are woven into the fabric of existence and shape the way objects in the world behave and interact. In a simplified way, this is what it means to say the world is number.

However, the same Pythagoreans who were solving mathematical problems to demonstrate that the world is number stumbled across a disturbing truth: there were things in the world that could not be expressed as a ratio of numbers. This discovery was so shocking that one person, Hippasus of Metapontum, was suspected of "drowning at sea" as a result. One of the most famous inventions, the Pythagorean Theorem, could be used to calculate one side of a right triangle—a triangle in which one of the angles is 90 degrees—if the other two sides are known. The theorem states that for all right triangles with shorter sides a and b and hypotenuse c (the longest side, opposite the 90 degree angle), $a^2 + b^2 = c^2$. For example, if know the two sides of a right triangle are 3 and 4, we can prove that the hypotenuse is 5.

$$3^2 + 4^2 = 5^2$$
$$9 + 16 = 25$$
$$25 = 25$$

Students, mathematicians, and scientists have used this simple and powerful formula for more than 2,500 years, but the Pythagoreans soon learned that this same formula would reveal their own downfall. If we turn to another simple example, let us assume that the two shorter sides are 1 and we want to calculate the hypotenuse.

$$1^2 + 1^2 = c^2$$
$$1 + 1 = c^2$$
$$2 = c^2$$
$$\sqrt{2} = c$$

The Pythagoreans were probably not terribly surprised to see this result and then got to work determining what number multiplied by itself equals 2. Through an iterative process, we know that 1.5 is too big, 1.4 is too small, and so on. The eager Pythagoreans were most certainly patient during this process, adding one decimal place at a time, until they faced a horrifying conclusion: the decimal never ends or repeats! How was this possible? If the world is numbers and ratios of numbers (to use modern language, if the world consists of rational numbers), then the existence of a number like $\sqrt{2}$ represented a crisis of faith.

This was the discovery of irrational numbers. The worst part for the Pythagoreans was this was not a one-off event: as it turns out, most numbers do not have a rational number as a square root. Arithmetically, we can determine the perfect squares by adding odd numbers sequentially:

1	$= 1$	and $\sqrt{1} = 1$
$1 + 3$	$= 4$	and $\sqrt{4} = 2$
$1 + 3 + 5$	$= 9$	and $\sqrt{9} = 3$
$1 + 3 + 5 + 7$	$= 16$	and $\sqrt{16} = 4$

The bad news for the Pythagoreans was this means that all the other numbers, $\sqrt{2}$, $\sqrt{3}$, $\sqrt{5}$, $\sqrt{6}$, $\sqrt{7}$, $\sqrt{8}$, $\sqrt{10}$, $\sqrt{11}$, $\sqrt{12}$, $\sqrt{13}$, $\sqrt{14}$, and $\sqrt{15}$, are irrational numbers.

There are two ways to consider this predicament. First, we can accept that irrational numbers like $\sqrt{2}$ are real and part of how the world operates (the world is still number), and then expand our understanding of numbers and mathematics to accommodate irrational numbers. Second, we can conclude that numbers and mathematics have a limited ability to correspond to everything in the world of space and time (the world is not number), just as language can never account for everything in existence, no matter how far we expand our vocabulary. After all, the decision to calculate the hypotenuse of a right triangle is a human invention, not something woven in to the fabric of existence, and the only way to do this is to arbitrarily impose a numbering system onto the triangle. Perfect triangles do not exist in the world. After all, a triangle still has three sides. If someone asks us how many cookies we want, we would never say $\sqrt{2}$.

However, the trouble did not end here. Numbers like $\sqrt{2}$, $\sqrt{3}$, and $\sqrt{5}$ would turn out to be manageable in Cantor's world of infinity because they are generated from natural numbers like 2, 3, and 5 (more on this later). However, the Pythagoreans would discover another type of irrational numbers that would really call into question their claim that the world is number. For example, if we consider the ratio of the circumference of a circle to its diameter, we get the famous irrational number π, which continues forever after the decimal without ending or repeating. The discovery of irrational numbers shook up the intellectual world of ancient Greece, and irrational numbers will be important for our analysis of infinity. Do irrational numbers "exist"? How many irrational numbers are there? Are there more irrational numbers than rational numbers? If we have to use an iterative process to generate irrational numbers one decimal place at a time, do we ever reach actual infinity and thus make the transition from a rational number to an irrational number? Do such questions make sense? The important point is that the crisis of irrational numbers did not arise until mathematicians attempted to impose arithmetic (the mathematics of quantity) onto geometry (the mathematics of the continuum), that is, to "arithmetize" the continuum. Therefore, before we get too perplexed by these results, we should first ask whether the results are a function of our numbering system and methodology.

Plato

Legend has it that engraved above the door of Plato's Academy were the famous words, "Let no one ignorant of geometry enter." This would suggest that mathematics was important for Plato. The ancient Greeks put more emphasis on geometry, due to the problem of irrational numbers in arithmetic, but the pendulum swung the other way with modern mathematics, back to arithmetic, with the goal of "arithmetizing" the continuum and replacing intuition with logic and rigorous definitions. With the end of infinity, I predict the pendulum will swing back to geometry.

Plato had important things to say about mathematics, but his dialogs make limited reference to mathematical problems. There is an episode in the *Meno* of the slave boy "recalling" the length of the hypotenuse of a right triangle ($\sqrt{2}$), but this problem was used to demonstrate Plato's theory of recollection, not to make a contribution to mathematics. In the *Philebus*, Plato raises the apeiron (unlimited) and the peras (limit): "As the first I count the unlimited, limit as the second, afterwards in third place comes the being which is mixed and generated out of those two. And no mistake is made if the cause of this mixture and generation is counted as number four?"[2] The reason Plato is important for the history of mathematics is because he provides a philosophical foundation for mathematics by showing how it fits into his theory of knowledge (epistemology).

The aim of Plato, like most philosophers, was universal truth, not solving particular problems. As a student of Parmenides, Plato agreed that our fleeting perceptions in many ways are not real (reliable) and therefore did not provide a solid foundation for philosophical truth, which raises an important point. If we cannot rely on our perceptions in our pursuit of philosophical truth, Plato argued we should turn to reason and intellect. After all, when we make a claim about ethical behavior, we are not relying only on our perceptions. Plato does not claim that our perceptions are devoid of truth or that we cannot use our perceptions to make true conclusions, but by raising philosophical truth to a higher level, he suggests that there are degrees of truth.

[2] Plato, *Philebus*, 27c.

In the *Republic*, Plato uses the celebrated simile of the Line to track the development of the human mind from ignorance to knowledge. The lowest kind of knowledge he calls Opinion, which has two levels and is primarily concerned with perceptions of the visible world. First, Low Opinion or images relies on second hand perceptions, such as people in a Greek village gossiping about the results of the Olympics in Athens. No one in the village attended the events in person but many of them heard versions of the events from family or friends who attended the events. Plato likens this to secondary images, such as, "shadows, then reflections in water and in all close-packed, smooth, and shiny materials, and everything of that sort, if you understand."[3] Obviously, it is impossible to make the transition from ignorance to wisdom based on this type of knowledge.

Second, High Opinion relies on direct experience of the objects or events that we had images of in Low Opinion, such as a person who attended the Olympics in Athens. "In the other subsection of the visible, put the originals of these images, namely, the animals around us, all the plants, and the whole class of manufactured things."[4] Granted, two people can watch the same event and reach two different conclusions, such as which athlete had better form or how the crowds reacted, which demonstrates a weakness of this methodology for attaining truth, but this direct exposure to particular things or events is more reliable and shapes our perceptions in ways that makes them more capable and discerning of truth for future events. According to Plato, Opinion never rises to the level of scientific knowledge, even when collecting empirical evidence. This might sound surprising to modern scientists or empiricists, who argue that all truth is grounded in empirical perception. However, this idea was essential to Plato's philosophy, and modern philosophy has demonstrated that we need something other than pure empiricism to account for the vast complexity of human knowledge.

When we move from Opinion to Knowledge, we also have two levels, Low and High, but in this case we will begin with High Knowledge. Plato is best known for this theory of Forms or Ideas, which is the realm of High Knowledge, the ultimate aim of his philosophy: "it makes its way to a first principle that is not a hypothesis, proceeding from a hypothesis but without

[3] Plato, *Republic*, 509e.
[4] Ibid, 510a.

the images...using forms themselves and making its investigation through them."[5] We normally think of words like horse or justice as mere abstractions that help us communicate with each other and organize our thoughts. However, Plato believed these Forms or Ideas exist independently and eternally, apart from the human mind and the world of perception, and that we meditate on these Forms or Ideas with reason and intellect, not with perception. There are particular horses in the world, but they are in a constant flux, never permanent, which makes true knowledge of them impossible.

The Form or Idea of Horse is an exemplar that all the particular horses participate in but are not identical with, like ten different documents printed from the same computer file. The same goes for more abstract ideas like justice. We can see particular instances of justice in the daily interactions of people, but the eternal Form or Idea of Justice is the eternal exemplar that defines what Justice is and sets the standard for where we should aim, which cannot depend on the whims of individual people or societies over time. The important point for Plato was that true knowledge had to be eternal and unchanging, not fleeting and changing, which is why it has to exist in a non-material realm, which is the realm of reason and intellect, not perception. Plato's theory of Forms or Ideas merits more study and is not as crazy as it might sound at first glance, but the important point for this book is to contrast it with what Plato considers Lower Knowledge.

Plato said the realm of Lower Knowledge is mathematics, which is surprising from someone who admonished people to study geometry as a prerequisite for entering his Academy. Plato considered the abstract nature of mathematics a good model for High Knowledge thinking, a training ground, so to speak, but it fell short of the full glory of High Knowledge. Plato considered the objects of mathematics (numbers and shapes) intermediary objects, between High Opinion and High Knowledge, because the problems of mathematics move from hypothesis to conclusion in this world, not to first principles, which Plato considered the pinnacle of truth. "These figures that they make and draw, of which shadows and reflections in water are images, they now in turn use as images, in seeking to see those others themselves that one cannot see except by means of thought."[6]

[5] Ibid, 510b.
[6] Ibid, 510e.

Aristotle explained the difference well. "Further, besides sensible things and Forms he [Plato] says there are the objects of mathematics, which occupy an intermediate position, differing from sensible things in being eternal and unchangeable, from Forms in that there are many alike, while the Form itself is in each case unique."[7] In other words, the objects of mathematics (numbers and shapes) are intelligible particulars, not sensible particulars, and not intelligible Forms or Ideas. For example, when we use the power of reason to reflect on equine-like animals of the same kind or species, we move vertically to the Form or Idea of Horse, a unique destination. In the case of mathematics, however, we use numbers and shapes to solve problems in the world of perception, such as adding apples or calculating the side of a right triangle.

The advantage of studying Plato is that, rather than take the objects of mathematics for granted, like the pre-Socratics and many people today, Plato attempts to explain what mathematics is and how it fits into his theory of knowledge. This important step, as we will see, is critical for understanding infinity because we get our first inkling that mathematics is distinct from other activity that relies on reason. For example, the process of abstraction that results in the word horse is not the same process that results in the number 2. Numbers are not properties or parts of the world in the same way that horses are, precisely because, as Plato noted, the object of mathematics are intermediary objects. The number 2 can apply to two horses, two apples, or one horse and one apple taken together, but the word horse applies only to horses and the word apple applies only to apples. This might seem like a distinction without a difference, but as our analysis continues, we will see that the consequences are important. To wit, as we enter the Platonic realm or the mind of God—yes, we have no numbers.

Schopenhauer

Normally, Aristotle would be the next stop for our historical analysis of mathematics and infinity, but I decided to dedicate a chapter to him and will therefore shift to a modern philosopher for the rest of this chapter. Normally, when we shift from ancient to modern philosophy, Kant would be the go-to philosopher for a discussion of mathematics, but I have opted to

[7] Aristotle, *Metaphysics*, 987b 14-18.

highlight an overlooked philosopher who had important things to say, often with greater clarity, precision, and insight. Granted, Schopenhauer in many ways rehashed the philosophy of Kant, but he said more about mathematics and, like Plato, provided better insights about how mathematics fits into a theory of knowledge.

Like Plato, Schopenhauer was interested in philosophical truth, but rather than look at knowledge only vertically (Plato's four degrees of knowledge), he also looked at truth in terms of categories or grounds. For example, Schopenhauer would argue that the following four statements are all true but in importantly different ways:

1. If you drop an apple, it will fall.
2. If all A are B and all B are C, then all A are C.
3. 1 + 1 = 2.
4. If Caesar crosses the Rubicon, Rome will descend into civil war.

All of these statements are true but they are true in different ways. Schopenhauer concluded that there are four grounds of knowledge, what he called the fourfold root of the principle of sufficient reason. This means that for every true statement, we can put it into one of only four truth buckets. Each true statement fits into one of the four buckets and not into more than one bucket.

The first ground of knowledge, from example 1, is the world of perception, the world of causally interacting objects—the ground of becoming. This is the world of science and empiricism. "Thus the law of causality is the regulator of the *changes* undergone in time by objects of external *experience*; but all these are material."[8] Unlike Plato, Schopenhauer considered this a valid ground of knowledge and worthy of our pursuit. The second ground of knowledge, from example 2, is the realm of reason—the ground of knowing or logic. "Reason...therefore, has absolutely no *material*, but only a *formal*, content which is the substance of logic; and so this contains mere forms and rules for the operations of thought."[9] There are conflicting arguments about whether logic provides us new knowledge, but logical statements are nonetheless true.

[8] Schopenhauer, Arthur, *On the Fourfold Root of the Principle of Sufficient Reason*, 56.
[9] Ibid, 171.

The third ground of knowledge, from example 3, is the realm of mathematics—the ground of being, in particular, our intuitive understanding of space (geometry) and time (arithmetic). "Therefore, neither the understanding nor the faculty of reason by means of mere concepts is capable of grasping them, but they are made intelligible to us simply and solely by means of pure intuition a priori."[10] The fourth ground of knowledge, from example 4, is the realm of inner sense—the ground of willing, acting, and motivation. "The result of this is the important proposition: *motivation is causality seen from within.*"[11]

If we consider these four grounds of knowledge, there are two for the inner world and two for the outer world, just as there are two for permanence and two for change.

Inner World	Logic	Will
Outer World	Mathematics	Science
	Permanence	Change

As we can see, the ground of mathematics is about identifying permanence in the outer world, which is consistent with Plato's claim of mathematics as applying to the world of perception, not to the world of Forms or Ideas. (Schopenhauer's model explains why mathematics cannot be reduced to logic—it is a unique ground of knowledge—which will be important as our analysis continues.) The ground of science is also in the outer world, which raises an interesting question: how do we distinguish mathematics from science? If we make scientific conclusions about the outer world by empirical observation and collecting data, that is, with perception, with the goal of discerning causal patterns to attain truth, how do we discern permanent mathematical truths if nothing in the world is permanent? The answer provided by Schopenhauer is pure intuitions—intuitions of space and time. According to Schopenhauer, there are three a priori forms of the intellect—space, time, and causality. What this means is beyond the scope of this book, but the important point is that science addresses the a priori form of causality and mathematics addresses the a priori forms of space and time.

[10] Ibid, 194.
[11] Ibid, 214.

According to Schopenhauer, "What distinguishes this class of representations, in which time and space are *pure intuitions*, from the first, in which they are *sensuously perceived*...is matter."[12] Matter and causality go hand in hand, but if we remove matter from the equation it allows us to think about space in terms of position (geometry) and time in terms of succession (arithmetic) without the burden of matter or causality. Beginning with space, our ability to abstract ideal triangles from the world of causality and know truths about them, such as the fact that the inside angles of all triangles add up to 180 degrees, is proof that this ground of truth rests on pure intuition, not on concepts. The mere concept of a triangle tells us nothing about the properties of triangles.

With geometry, the axioms of Euclid are grounded on pure intuition, such as the claim that parallel lines never cross. An analysis of the concept of line will never generate this truth. We can use the axioms to derive other geometric truths in a deductive way, but the axioms themselves cannot be proven. The axioms provide the bedrock, and it would be a waste of time to prove what cannot be proven. In logic, on the other hand, the fundamental units of analysis that are used in syllogisms are concepts, not pure intuitions, which is why mathematics cannot be reduced to logic. Unlike the world of causality, in which we need repeated observations over time to discern patterns, the truths of geometry are pure intuitions because they can be identified and understood with one observation.

With arithmetic, the idea is similar but less complicated. If one day gives rise to another day and one year gives rise to another year, then if we abstract the process with numbers, absent the causality of days and years (planets rotating and orbiting), we have arithmetic. Rather than count 1 for this year and wait for the next year to add 1 to reach 2 (waiting for the causality of the earth's orbit around the sun), we can abstract this process and solve problems now and into the future without limits. For example, if I were to go back to reconstruct the life of a king, I could use public records to calculate the number of years he lived (1 + 1 + 1 ... = 78). As Schopenhauer noted, arithmetic is nothing but methodological abbreviations of counting, and each number presupposes the previous numbers. We cannot understand 10 unless we understand the 9 numbers that precede it. As I will address later, once we have

[12] Ibid, 193.

established the natural numbers (1, 2, 3, ...), we can use mathematical operations (addition, subtraction, multiplication, division) to expand the natural numbers as far as we need to solve problems.

Schopenhauer wrote before Cantor, but my guess is he would have been critical of Cantor's leap to actual infinity by highlighting that Cantor reached his conclusion by blending the intuitions of space (geometry) and time (arithmetic) in an unjustified way. Schopenhauer probably would have been equally critical of Cantor's attempt to reduce mathematics to logic because mathematics and logic are two distinct grounds of knowledge. Schopenhauer admitted that there was no theoretical limit to the process of division, but he probably would have balked at the idea of using this methodology to propose the metaphysical existence of actual infinity or to argue that something other than intuition provides the foundation for this knowledge.

One of the most important trends of modern mathematical philosophy has been the attempt to eliminate Kant's and Schopenhauer's intuitions by "arithmetizing" the continuum and reducing mathematics to logic and rigorous definitions. Admittedly, intuition can be a vague idea that is easy to abuse—people often equate "feeling lucky" with intuition—but the best way to arrive at a conclusion is to look at the evidence. For example, if mathematical intuition is real, then we would see infants demonstrating an ability to recognize mathematical truths without the benefit of an adult mind that can collect empirical data over time to identify patterns and reach conclusions. In a study by Karen Wynn published in *Nature*, she found that babies as young as five months old could calculate the results of simple arithmetic on small numbers of items, to include knowing when the problems are done correctly or incorrectly. This means babies understand (the category of) quantity before they understand the names of the numbers, and it demonstrates that babies have an innate sense for mathematics that cannot be credited to their understanding numbers or mathematics.

Conclusion

The first takeaway from this chapter is that understanding mathematics and infinity has nothing to do with solving mathematical problems. We use philosophical not mathematical thinking to understand what mathematics is and how it fits into a theory of knowledge. We do not need philosophy to solve mathematical problems (I solved Calculus problems in college without understanding the philosophy of mathematics), but we should not allow mathematical problems to take us down untethered journeys, like the idea of infinite divisibility, without a strong foundation in philosophy.

The second takeaway is that some of the most important philosophers have showed that mathematics is a unique ground of knowledge that cannot be reduced to logic. Just as we should be critical of any attempt to reduce science to logic, such as the rationalist philosophy of Spinoza, we should be critical of any attempts to reduce mathematics to logic. The movement of reducing mathematics to logic began with Frege and Russell, but it did not end well. Gödel proved that such a reduction was impossible, but he could have reached the same conclusion by reading the great philosophers.

Chapter 2: Aristotle

Philosophers tend to lean toward Plato or Aristotle—the great divide. No one is indifferent, no middle ground, and many friendships have ended as a result. Wars have been fought over the implications of their ideas. My own theory is that this preference is hard-wired, in the same way that personality traits are hard-wired, but the truth is they had more in common than most people are willing to admit. Plato and Aristotle framed many of the most important philosophical debates that are still with us today (metaphysics, epistemology, ethics, politics, and so on), and any study of philosophy is incomplete without an understanding their ideas and their legacy.

One particular issue was a notable source of conflict—Plato's theory of Forms or Ideas—and Aristotle went to great lengths to discredit Plato's theory. "Again, it would seem impossible that the substance and that of which it is the substance should exist apart; how, therefore, could the Ideas, being the substances of things, exist apart?"[13] If everything Aristotle said about Plato's theory is accurate, then it hard to imagine how anyone could accept it; but there are good reasons, like Plato's own writings, to conclude that Aristotle might have been unfair. Either way, Aristotle's response to Plato's theory of Forms or Ideas is important for mathematics and infinity.

If we recall, Plato believed that the pursuit of truth transcended the world of perception. We could perceive material objects like horses, but if we want to gain true knowledge of horses we have to shift our attention from the horses of perception to the Form or Idea of Horse, the eternal exemplar that

[13] Aristotle, *Metaphysics*, 991b 1-3.

we contemplate with reason and the intellect. This Form or Idea of Horse is eternal and unchanging, which is what Plato said was required for truth. Aristotle thought this was absurd and saw no reason to postulate a parallel realm of immaterial Forms or Ideas to gain truth about the world. After all, if these Forms or Ideas were not of this world and could not be perceived, how did we gain knowledge of them? Do we have special senses for Forms or Ideas? How did horses in the world of perception participate in or interact with the Form or Idea of Horse if they are immaterial? Aristotle agreed with Plato that truth was not to be found in the realm of perpetual change, but he insisted that we could look to the material objects in the world of perception to discover their unchanging truth or essence. The essence of a horse could be found within actual horses, not in an immaterial realm of Forms or Ideas. Today, we might point to the DNA of horses to identify a foundation for the horse that remains stable as individual horses grow, decay, and ultimately die.

Aristotle's argument that we can gain truth via perception had important consequences for philosophy and science, but he understood that reason is actively involved in the quest for truth and is not merely a passive recipient of perceptions. If Aristotle is right and we eliminate Plato's realm of Forms or Ideas, then we also must eliminate Plato's objects of mathematics (numbers and shapes) as intermediary, immaterial objects between perceptions and the Forms or Ideas. Thus, Aristotle will face the challenge of grounding mathematics in the world of perception with the assistance of reason. Aristotle made a sophisticated argument for the First Cause or the Prime Mover, but this was not the realm of numbers, shapes, Forms, or Ideas.

To start moving toward Aristotle's philosophy of mathematics, we should consider his view of definitions (essences) and how definitions apply to species, such as humans, not to individuals. For example, if I want to understand my own essence, I would do so by thinking of myself as a human, not as my individual self. Even the great Aristotle did not have a unique essence. If we want to define human, we have to include the genus, which is the kind under which the species fall, and the differentia, which characterizes the species within that genus. In his most celebrated example, Aristotle defined human as an animal (genus) with a capacity for reason (differentia)—the rational animal. With this, Aristotle concluded that reason was essential to our nature and distinguished us from all the other animals. With this methodology, there is no need to postulate a parallel world of immaterial Forms or Ideas where a Rational Animal exemplar exists.

As we apply this methodology to the other things we see in the world of perception, we eventually start to form a cohesive and consistent web or hierarchy of concepts that is a rational reflection of the world of perception that allows for logic, which I will address in more detail later. Aristotle also postulated the existence of things we could not see, such as the First Cause or the Prime Mover, but he was able to make this intellectual leap only after a thorough study of the universe—the transition from physics to metaphysics.

The Categories

If we accept that the world of perception is a source of truth or knowledge and that the essence of an object is within the object, not in the Platonic realm of Forms or Ideas, we can focus our efforts on studying the world in a scientific and philosophical way to gain the truth that Plato sought. As we do this and conceptualize the world, we soon realize that our concepts are not a mere plurality of unrelated things that we toss into a bucket; rather, many concepts are related to each other horizontally and vertically—hierarchically. For example, we can study animals around the world and identify species with similar features. The process involves some trial and error, such as zooming too far out ("organism") or zooming too far in ("fruit fly"), but we eventually get it right. Our observations might not always be accurate, but a careful review of the physiology and DNA of these species will show that they are related to each other in ways that they are not related to other species. For example, we might identify species like tigers, lions, leopards, jaguars, and snow leopards in the following way.

Genus Pantera					
Species	Tiger	Lion	Leopard	Jaguar	Snow Leopard

As we learned in biology, all living organisms fit into the framework of kingdom, phylum, class, order, family, genus, and species, which is a human creation that serves an epistemological function that "works" but also accurately reflects reality. In theory, we could revise this model but the marginal gains probably would not be worth the effort. Beneath the tiger species, we can identify sub-species of tigers, and so on, but the process will reach a natural limit, especially in the case of animals. The process does not continue ad infinitum, as it often imagined in the case of numbers.

This framework helps us understand that in addition to identifying the essence of individual objects to arrive at a clear concept, we also understand how these species relate to each other hierarchically. Granted, "cats" per se do no exist, only individual species of cats, but there should be no need to justify the existence of abstract concepts like "cat" because it often suffices for a given context, such as pets. "Do you like cats?" This same idea applies to man-made concepts.

Furniture				
Chair	Couch	Table	Bed	Dresser

Chairs do not exist in the same way that cats exist (we can imagine an alien race with no need for furniture), but once we establish a history of using and producing furniture, we can start talking about the essence of a chair for humans, at a minimum to not confuse it with other types of furniture. (Needless to say, Aristotle would scoff at the suggestion of the Form or Idea of a Chair in an immaterial realm serving as the exemplar for chairs in the world.) There are many types of chairs—desk, kitchen, stool, office, bench, and so on. This process can continue as well, such that a store could specialize in office chairs, with different types of office chairs (color, shape, design, material, and so on) based on age, size, gender, weight, income, and so on. As was the case with animals, this process can continue, but although there is no theoretical limit to this process, because chairs are an artificial creation that is limited only by our imagination, there are natural limits because it would not make practical sense to have 25,278,792 different types of office chairs in a retail store or catalog.

Just as we can work our way down, such as from genus to species (analysis), we can also work our way up (synthesis), such as from species to kingdom. As we map out the various conceptual networks in the world, such as animals and furniture, and move up to higher levels of abstraction, we also reach natural limits and arrive at multiple but limited abstract concepts or modes of being outside the mind that cannot be further combined through synthesis: "for the greater class is predicated of the lesser, so that all the differentiae of the predicate will be differentiae also of the subject."[14]

According to Aristotle, the abstract concepts or modes of being at the end of this road are called categories, of which he lists 10: substance, quantity, quality, relation, place, time, position, state, action, and affection. We cannot combine any of these ten categories to arrive at higher abstractions or modes of being. It is not easy to understand how these 10 categories (in one place he lists only eight) are abstractions that also have metaphysical significance (they are supposed to bridge the gap between logic and metaphysics), and other philosophers like Kant would invoke them in ways that made life easier for no one. Aristotle was not always consistent or clear about the categories, but the important point for Aristotle's theory of mathematics and infinity was that he grounded mathematics in the category of quantity.

As we observe the world, we perceive a plurality of objects. Importantly, Aristotle concluded that this plurality of objects is real and a source of knowledge, not an illusion or a source of ignorance. Tigers and chairs are metaphysically distinct entities that can be counted. In a world without quantity (a plurality of objects), numbers and mathematics would never arise. Whether we observe an apple orchard or an Olympic stadium, there will be a quantitative piece of the observations that cannot be further abstracted via synthesis. For example, we might see 100 apple trees or 10,000 people in the audience, but we also might see 10,000 apples or 100 athletes. The numbers 1, 2, 3, ... should be clear and unambiguous (more on this later), but it is not always clear when they apply. Is the person standing in front of me 1 (body), 2 (body and mind), 3 (body, mind, and soul), or the billions of atoms that make up his body? If 2 clouds collide and merge (1 + 1), do we have 1 cloud or 2? My cat might have 10 kittens or 1 litter of kittens. In other words, the use of numbers has both natural and arbitrary elements that depend on the context and our choices—on reality and reason.

Different numbers can apply to the same object depending on the context. In other words, specific numbers cannot be predicated of specific objects, which has important implications for mathematics and infinity that I will address in more detail later. Likewise, the individual numbers cannot

[14] Aristotle, *Categories*, 1b 21-23.

be arranged hierarchically in the same way that animals or furniture are. We count the numbers in a sequential order (1, 2, 3, ...), but we can also derive numbers from other numbers (1 + 1 = 2). As a result, given that Aristotle rejected the Platonic realm of Forms or Ideas and given that individual numbers are not substances or do not have essences (more on this later), Aristotle would ground numbers in the plurality of objects in the world (the category of quantity), which, as we will see in the next section, will have important consequences for his understanding of infinity.

Potential Infinity

Aristotle's most important conclusion about infinity was that infinity existed only potentially, not in actuality, which was an idea with lasting consequences for the history of mathematics and philosophy, and is the position I will argue for in this book. "Nor can number taken in abstraction be infinite, for number or that which has number is numerable."[15] (Of interest, Cantor would claim that it makes sense to talk about numerable and innumerable infinite sets.) Some have argued that Aristotle's view of potential infinity prevented the ancient Greeks from inventing the Calculus, until Leibnitz and Newton invented it about 2,000 year later, but I will prove that this claim is false because infinity is not required for the Calculus.

Because the category of quantity for Aristotle is grounded in counting things in the world of perception, not in an intermediary realm between the world of perception and the world of Forms or Ideas, the idea of actual infinity is a contradiction in terms because there is no theoretical end to the counting process. The idea of an actually infinite set, such as the real numbers between 0 and 1, is impossible after we clearly define our terms and fully grasp the implications. No matter how high we count, we can always add another number, and the idea of a completed infinite set of numbers with extension on a continuum violates the law of contradiction.

Aristotle recognized two types of quantity: discrete and continuous. "Instances of discrete quantities are number and speech; of continuous, lines, surfaces, solids, and, besides these, time and place."[16] This difference is important: discrete quantities are metaphysically distinct entities that can be

[15] Aristotle, *Metaphysics*, 204b 6-8.
[16] Aristotle, *Categories*, 4b 23-24.

counted by different people with identical results, regardless of context, like marbles in your pocket. Continuous quantities, on the other hand, are continuums that require us to arbitrarily impose numbering systems onto them, which can change for different people and different situations. There is nothing inherently quantitative about a continuum, aside from calling a continuum 1, such as one side of a triangle. Saying a triangle has 3 sides (quantity) is not the same thing as saying one side of a triangle has a length of 3 (continuum) because whereas a triangle has three sides regardless of how we perceive them, the side of one triangle having a length of 3 depends on our arbitrarily imposing a unit of measure onto the triangle.

Because Aristotle thought of discrete counting as applying to metaphysically distinct objects, the idea of actual infinity was impossible because if there were an infinite quantity of material objects in the world, the universe would be infinitely large, which was the same argument made by Zeno. (Aristotle would also allow for the counting of immaterial objects, like ideas or dreams, but he would still reject the idea of an actual infinite set of ideas or dreams.) Even the modern Cantor camp would agree that discretely counting discrete objects never reaches actual infinity. Infinite sets must be thought of as completed wholes.

In the case of continuous quantities, what Aristotle called the infinity of division, he granted that infinity had more of a potential reality because we could continue to divide continuums ad infinitum without running the risk of an infinitely large universe. A continuum does not grow larger as we sliced it into smaller parts. However, the dividing process will never result in dimensionless ranges, which is required to achieve an infinite set. The fact that we can see a line segment with our own lying eyes makes it more palatable for the infinity of division, but this is nothing more than a trick of the eye that leads us into error. Rather than argue that the side of a triangle could be infinite in length, Aristotle argued that there was no theoretical limit to how many times the side could be divided into smaller slices, which is really two ways of saying the same thing. For example, suppose we take a line with an arbitrary length of 1 and divide it into 2 equal parts. In this case, we double the number of units but make each unit half as long, which means the line still has a length of 1. We can continue this process ad infinitum:

10 parts with length of 1/10 = 1
100 parts with length of 1/100 = 1
1,000 parts with length of 1/1,000 = 1
10,000 part with length of 1/10,000 = 1
...
100,000,000,000 parts with length of 1/100,000,000,000 = 1
...

No matter how many times we divide a finite continuum into smaller units, the total length of the units does not change because as the quantity of units increases, the range of the individual units shrinks proportionally. We can continue this process ad infinitum to achieve whatever level of accuracy or precision is required. However, just as we can never reach actual infinity by counting discrete objects, we can never reach actual infinity by dividing the continuum. There is no actual infinity of quantity waiting behind the veil of perception for us to discover with an intellectual leap. In fact, the idea of achieving actual infinity would amount to the ranges of the individual units shrinking to a length of 0, which would result in a singularity implosion and the disappearance of the line or continuum.

I believe Aristotle's conclusion about potential infinity was correct. We can demonstrate that he was correct and that we gain nothing by introducing the idea of actual infinity, but his language in two areas planted seeds for the idea of actual infinity that he opposed. First, he did not draw enough attention to the fact that the two types of counting are really the same. "Now, as we have seen, magnitude is not actually infinite. But by division it is infinite."[17] Whether we are counting discrete objects in the world of perception or counting points on a line resulting from division, the counting is the same. If actual infinity is not possible with the infinity of addition, it is not possible with the infinity of division. We can divide a line into 2 parts, 3 parts, 4 parts, and so no, ad infinitum. In other words, although Aristotle fully abstracted numbers in the case of division counting, he did not do so in the case of discrete counting, which is a problem that continues today. When we properly abstract numbers, counting is counting, and neither counting nor axiomatic decree will ever take us to actual infinity.

[17] Aristotle, *Physics*, 206a 16-17.

Second, Aristotle did not draw enough attention to the fact that the mathematics of the continuum is an arbitrary creation of the human mind. There is nothing inherently quantitative about the continuum and no reason to believe that a line consists of infinitely many points, that a plane consists of infinitely many lines, or that an object consists of infinitely many planes. This means that when we arbitrarily impose numbers onto a continuum, we should not allow the continuum, in turn, to generate truths about numbers that the numbers themselves do not generate. For example, the claim that the set of rational numbers leaves "gaps" on a continuum does not mean that an infinite set of real numbers must exist to "fill the gaps." Dimensionless points can never be stacked or lined up to achieve extension ($0 + 0 = 0$), and the requirement for points with dimension to achieve extension makes actual infinity impossible along a finite continuum.

Divine Ideas

One of the most important questions in the infinity debate is what purpose does actual infinity serve? What is gained with its existence? What is lost if its existence is rejected or discredited? Why are some people so passionate about it? Just as important, what does actual infinity mean? After all, the universe is not infinitely large, has not been around for an infinite time, and does not contain an infinite number of particles. In fact, in the context of infinity, the total number of atoms in the universe is a trivially tiny number, and one billion years is a trivially tiny fraction of a blink of an eye in the context of infinity. As we dig deeper on this topic we will see that the debate about infinity is really about much more than numbers. In fact, the idea of infinity gets to the heart of the most abstract debates in philosophy and theology, and the direction we go will have important consequences for philosophy, mathematics, and beyond.

To get us moving in the right direction, if we consider tangible objects in the world, such as humans or trees, their existence raises important questions. In each case we can ask, what made it possible? As individuals, our parents made us possible, their parents (our grandparents) made them possible, and so on. An acorn made the oak tree possible, and so on. We can speculate on how far back this causal chain goes or whether the idea of an infinite regression is possible, but one thing is clear: each object in the world is not capable of generating itself out of nothing. That is, every object in the world has a contingent existence that depends on something else for its existence. If every object in the universe is contingent and every contingent object depends on something else for its existence, so the thinking goes, there must be something that is not contingent, whose existence does not depend on something else—something to bring contingent objects into existence and put the whole thing in motion.

Shifting from the seeds of theology back to our discussion of numbers and infinity, is this also the case with numbers? Do numbers have a contingent existence that depends on something else? Do we need a non-contingent thing to account for the existence of numbers or can they be created *ex nihilo* by contingent beings, like humans? Granted, in this context, numbers are contingent to the extent that humans are contingent. Or, as Pythagoras claimed, are numbers non-contingent and the source of everything we see? Humans make arbitrary numbering systems to suit their own needs and limitations, such as base-10, place-value numbers, and this numbering system was absent for most of history, but perhaps we are merely symbolizing something that exists at a deeper level? Just as a rock does not suddenly exist when we say the word "rock," perhaps the numbers do not suddenly exist when we say them.

Aristotle argued that the category of quantity provides the foundation for mathematics; the perception of quantity gives rise to number, and we use numbers to count the quantity of objects in the world (pizzas) or to divide these objects into smaller pieces (slices of pizza) that we can count as well. Reason abstracts the category of quantity from the world of perception, not from a transcendental world of Forms or Ideas, but we can also imagine a non-contingent being creating a plurality of objects, which is how the category of quantity can be thought of as bridging the gap between logic and metaphysics. However, we do not require metaphysics to account for the existence of numbers. In one case we might predicate 1 of a human but later predicate 2 of the same human if he is with another human, or a much larger number if we consider him as a collection of atoms. A baby could eat two pieces of candy even if he has no grasp of the number 2.

To use the language of Aristotle, number is not a substance and does not have an essence. There are no essential differences among the numbers (1,

2, 3, ...), and one number can generate another number. "Again, if the Forms are numbers, how can they be causes? Is it because existing things are other numbers, e.g., one number is man, another is Socrates, another Callias?"[18] We can define number in a general sense or for individual numbers, but this does not help us solve mathematical problems. A child can add 3 + 5 = 8 without a definition of number but a child cannot have a meaningful discussion about political science without precise definitions for the relevant terms. The number 1 does not differ from the number 2 in the same way that a dog differs from a cat. Even Plato denied number a place in the realm of the Forms and Ideas. For Plato, numbers are intermediary objects that solve problems in the world of perception. The numbers do not exist like the Forms or Ideas or take us beyond this world.

Perhaps this analysis misses the point? Perhaps Pythagoras was correct that the world is number? According to Plato, the Demiurge used the Forms or Ideas and the Platonic solids to fashion the world and impose order on the chaos. Plato does not explain the origin of the Demiurge or of the raw material of the world, but he tells us something about how it came to be: "it is a work of craft, modeled after that which is changeless and is grasped by a rational account, that is, by wisdom. Since these things are so, it follows by unquestionable necessity that this world is an image of something."[19]

Rather than think of Plato's Forms or Ideas as mere abstractions, we should think of them as the models or exemplars that were used to create the world. Consider a 3D printer. The quality of the final product depends on the quality of the computer file, the printer, and the printing materials (the 3D "ink"), but each product is fashioned by the same Form or Idea (the same digital file that sends instructions to the printer). Likewise, the products will age, break, and decay, but if we want to understand the essence of an object, we have to reverse engineer the flaws and decay to seek out the eternal Form or Idea that made it possible, so that we could recognize a new one if we saw it. We should not allow the aged, broken, or decayed objects taint our understanding of the truth.

In more modern thinking, the same idea applies. If the objects in the world appear to have a specific essence or nature, then the divinity or God

[18] Aristotle, *Metaphysics* 991b, 9-11.
[19] Plato, *Timaeus*, 29a 8 - 29b 2.

must have used some kind of design, exemplar, or pattern to make them. This would suggest that the divinity or God thinks about things, but thinking is a temporal activity—one thought followed by another thought—which is usually thought of as inconsistent with the idea of the divinity or God. Thus, some philosophers and theologians talk about divine ideas existing in the mind of the divinity or God, which is more consistent with Plato than with Aristotle because it explains how an immaterial exemplar could provide the foundation for a material object.

The divinity or God does not think about divine ideas the same way we understand thinking, but the divine ideas can be thought of as shaping creation, just as a novelist uses mental material to shape the way he writes a novel. Aristotle's First Cause or Prime Mover, however, did not create the world (the world is eternal) and thinks of nothing but itself and thus has no divine ideas. In this case, mathematics is nothing more than humans using arbitrary yet effective numbering systems to count objects and solve problems. Thus, it appears that some form of Platonism is required to talk about numbers and infinity having an existence apart from or independent of the material world, some type of nonmaterial realm that we cannot perceive.

Cantor was a religious man and thought about infinity in the context of God, so let us suppose that some sort of immaterial stuff (Forms, Ideas, divine ideas, algorithms, etc.) are used to fashion the world. The question is, is it possible or necessary for an infinite set of numbers to exist in this immaterial realm to provide a metaphysical foundation for our world and mathematics? The first way to fashion the world from this Platonic realm, in addition to the Forms and Ideas, is to use formulas or algorithms. For example, let us suppose that formulas like $E = mc^2$ and $F = ma$ are not mere human abstractions but the actual formulas or algorithms for how the universe works—the code of the universe. Just as a computer follows the directions of the written code, perhaps the universe operates according to these scientific formulas or algorithms, which would then raise questions about the origins of the formulas or algorithms. If so, then there is no need for an infinite set of numbers to exist in the immaterial or Platonic realm. The formulas or algorithms suffice, just as a computer programmer does not need to have a file with a list of all the numbers the code might use. The people using the code provide the numbers as needed, and there is no reason to limit this to base-10, place-value numbers.

Likewise, consider the Platonic solids (tetrahedron, cube, octahedron, dodecahedron, icosahedron). If these shapes are used to construct the objects of the world, the shapes alone suffice, just as a child can construct a castle with building blocks. We do not need numbers in the Platonic realm to construct geometric shapes. If we recall, irrational numbers often result from the ratios of the sides of shapes, such as the ratio of one side to the hypotenuse for a right triangle. There is no need for $\sqrt{2}$ to exist in the Platonic realm if the sides of a right triangle will generate it via the Pythagorean Theorem when a human mind imposes a numbering system. There is no need for π to exist in the Platonic realm if the ratio of the circumference to the diameter of the circle produces it, just as people in the ancient world could draw circles with no understanding of π. In other words, it is incorrect to say that π frequently appears in nature; the truth is that geometric shapes and ratios frequently appear in nature that just happen to be expressed arithmetically as π.

In short, the irony of the Cantor camp invoking Platonism to provide a metaphysical foundation for actual infinity is that Plato himself declared that numbers and shapes do not exist in the Platonic realm. The realm of mathematics is the human mind, our own algorithms, even if nature reveals deeper patterns that can be expressed mathematically. Numbers are human creations that solve problems in the world of perception, including realms that are too big or too small to be perceived by the human eye, and the nature of our base-10, place-value numbering system reflects the thought processes and limitations of the human mind. Just as whatever created the world did not need the word "cat" to produce cats, it did not need the number 3 to produce 3 objects.

Conclusion

Some of this analysis was admittedly abstract, especially the last section on divine ideas, but the ideas will become more tangible and concrete as we continue our analysis. The important point with Aristotle is that if you accept his philosophy, then actual infinity is impossible and in many ways meaningless. The Cantor camp disagrees with Aristotle on this point, which is why they resort to Platonism and an immaterial realm to provide a foundation for actual infinity, where the infinite sets of numbers supposedly exist in all their glory.

The irony of the Cantor camp invoking Plato is that Plato said mathematical objects (numbers and shapes) do not exist in the realm of Forms or Ideas and thus did not shape the creation of the world. Mathematics does not require a metaphysical foundation of actual infinity, and we cannot bring something into existence by definition or axiomatic decree. Whatever created the world with a plurality of objects (quantity) did not need numbers to do so. Plato admitted that mathematical objects (numbers and shapes) have an intermediary existence, but Plato would have disagreed with Cantor's claims about numbers or infinite sets existing in the realm of Forms or Ideas. Not to mention, Cantor claimed we can expand an infinite set by taking the power set of the infinite set (more on this later), which opens the door to an infinite regress of ever expanding infinite sets and no foundation where the process stops (or begins, depending on your perspective). This erroneous thinking is similar to asking what created a non-contingent divinity or God.

Chapter 3: Language & Logic

Language shapes how we think, in positive and negative ways. On the positive side, language allows us to refine our thoughts and communicate with others with precision and nuance, to include the use of metaphor, which could never be achieved in a world of grunts and gestures. It might not always matter whether someone describes our work as "excellent" or "outstanding," because both adjectives are in the "very good" bucket, but saying you "really like" rather than "love" someone can have disastrous consequences for your relationships. On the negative side, imprecise language or the misuse of language can lead to confusion or misunderstandings, even violence. How many wars have been fought and how many millions of people have died for invalid ideas? If people allow words they do not understand to shape how they understand the world, rather than using words to express their understanding of the world, then language will often be a barrier to clear thought and communication. Even worse, language will be used with the specific purpose of obfuscating the truth and manipulating people in nefarious ways.

One of the most important trends in modern philosophy has been the focus on language and logic, which is known as analytic philosophy and was promoted by philosophers like Frege, Russell, and Wittgenstein. The goal of this movement was to solve some of the most important problems of philosophy by clarifying the language we use, which resulted in many of the philosophers concluding that many of the problems of philosophy, such as metaphysics, are meaningless and not worthy of our attention. Most of this methodology focused on using clearly defined terms and logic to root out the confusion. For over 2,000 years, the subject-predicate syllogistic logic of Aristotle dominated philosophy (Russell claimed Aristotle's metaphysics was baked into the cake of his logic), but the analytic philosophers developed modern formal logic and other forms of logic to attempt to solve these problems. This resulted in a new focus on mathematics and attempts to reduce mathematics to logic, which ultimately did not succeed. In fact, logic itself would show us that such a reduction was impossible.

Although the analytic philosophers looked to the purity of mathematics for inspiration and some progress was made, these philosophers did not focus on the fact that mathematics is a language with its own flaws, to the extent that it is designed to meet the needs of the human mind. Most people take it for granted that we have a base-10, place-value numbering system that has its own structural issues and peculiar properties, such as the interesting properties of the number 9.

Just as Russell claimed that the idea of God is grounded in flawed thinking and bad language, as he tried to demonstrate with his logic, he did not seem to consider that his own faith in actual infinity might be grounded in flawed thinking and bad mathematics. He incorrectly assumed that we could use mathematics to solve the problems of mathematics (meta-mathematics), and apparently did not recognize that the argument about infinity is ultimately philosophical, not mathematical. "On this basis he [Cantor] was able to build up a most interesting mathematical theory of infinite numbers, thereby taking into the realm of exact logic a whole region formerly given over to mysticism and confusion."[20] After all, the only way to make the leap to actual infinity within set theory was decree—the Axiom of Infinity. It is never a good idea to assume the truth of something to prove that it is true, a logical error known as begging the question. This line of thinking would have merit if we could not solve mathematical problems without assuming the existence of actual infinity, but the truth is we can solve mathematical problems without infinity.

[20] Russell, Bertrand, *The History of Western Philosophy*, 830.

Place-Value Numbers & Letters

Language and mathematics are more similar than most people imagine. For example, consider the following passage:

> The boy returned from the pet store. He opened a plastic bag and poured the goldfish into the fishbowl. He opened the other plastic bag and poured the other goldfish into the same fishbowl.

Now let us consider this passage in the language of mathematics:

$$1 + 1 = 2$$

If we keep in mind that Aristotle said mathematics was the science of quantity, then mathematics is the language of quantity, except that in mathematics the only thing we have are quantities (nouns), such as numbers, and the things we do to them (verbs), such as add, subtract, multiply, or divide. Just as the power of physics depends on its ability to strip away everything in the world that is not quantitative, the power of mathematics depends on its ability to strip away everything in language that is not quantitative and to use abstract symbols with universal consent.

One of the challenges of learning Chinese is memorizing a unique symbol for each word. Some of the symbols are combinations of symbols, but for the most part you have to learn a new symbol for each word. Just as important, if you have never seen the symbol before, there is no systematic way to know how to pronounce it or to understand what it means. At first glance, this seems difficult and inefficient, which it is, but we also have to learn a unique word for most words in English. Root words, prefixes, and suffixes make this easier, but the process is not easy. Ask anyone who has tried to learn a foreign language. In terms of symbols and sounds, there is no theoretical limit to how a language is constructed, but there are two important things that limit language: first, the limits of humans to make sounds; and second, the human need for efficiency.

There are a finite number of clear sounds that we can use our mouths and tongues to articulate to other people and be understood on a consistent basis, which in turn means that we have to combine this limited quantity of primary sounds to pronounce the thousands of words we use. (If we never spoke to each other and only needed abstract symbols to communicate, our language could be very different.) To make the transition to writing, we need an alphabet, which uses individual letters or combinations of letters to represent the individual sounds we make with our mouths, meaning that sounds are the foundation of our written language. Granted, we can adjust the letters to make us pronounce sounds we are not used to pronouncing, but the sounds ultimately provide the foundation for the alphabets.

Not all cultures have the same sounds or same alphabets, but there are some surprising consistencies around the world and throughout time, such as the number of letters and the types of sounds. In the case of English we can use the 26 letters of the alphabet to construct tens of thousands of words, which takes us to efficiency. We could create more unique sounds or letters, but this would be a waste of time and mental effort. The limited number of sounds and letters undoubtedly place limits on language, but the efficiency gains outweigh the cost. Having the words "to," "too," and "two," might seem odd to an outsider, but these peculiarities make our language more manageable.

The same idea applies to numbers. In theory, we could have a unique number for each number, such that we would have to memorize 1,000 unique symbols for the first 1,000 numbers, but this would limit mathematics in significant ways. We would have to reach consensus for each unique symbol, such as 1,000,000 and 1,000,001. In the category of inefficient numbering systems, consider notches on a bone or Roman numerals. If our forays into quantity never go beyond 20 or 30, we could probably survive with making notches on a bone, such as keeping track of how many enemy warriors we have defeated. Two different warriors could compare the number of notches on their bones to see who was the best warrior. In the case of Roman numerals, they are more systematic but there is nothing intuitive about them and, most important, there is nothing built into the system to help us solve problems, such as V - IV = I or II x V = X.

Similar to the alphabet, the base-10, place-value numbering system is one of the greatest inventions of all time. Our numbers are base-10 because we use 10 numbers (probably because we have 10 fingers) and they are place-value because the value of a number depends on its place in the number. For example, a 1 in the first position is 1, a 1 in the second position is 10, a 1 in the third position is 100, and so on. On the other side of the decimal, a 1 in the first position is 1/10, a 1 in the second position is 1/100, a 1 in the third position is 1/1,000, and so on. Unlike words, the numerals do not tell us how to pronounce them. They are not phonetic. They are pure symbols. We could write seventeen plus twenty-four equals forty-one, but 17 + 24 = 41 is much better.

Just as we can use the 26 letters to construct tens of thousands of words, we can use 10 numbers (0–9) to construct numbers ad infinitum. We are not limited by the ability of our mouths to make sounds. Not only that, the numbering system itself helps us solve problems with impressive efficiency, just like a lever helps us move heavy objects. The base-10, place-value numbering system is a powerful tool. If two warriors wanted to compare the number of notches in their bones, they would have to sit down and compare the two bones by looking at each notch. If the tribe had a system for marking the notches a fixed distance apart, such as the width of a magic pebble, they could see the problem visually and count the extra notches on the top bone to see he as 12 more.

||||||||||||||||||||||||||||||||||||||||||||||||||

||||||||||||||||||||||||||||||||||||||

In the case of Roman numerals, the system is less effective because we could not work with or manipulate the letters to solve the problem. We have to do the problem in our head or use some other methodology and then write down the answer.

M - XXXVIII = XII

In the case of base-10, place-value numbers, however, the problem almost solves itself once you understand the notation and rules.

$$\begin{array}{r} {}^{4}5\,0^{10} \\ \underline{-38} \\ 12 \end{array}$$

The power of this system derives from the fact that we can solve massively large problems by merely adding, subtracting, multiplying, or dividing small numbers, 0 - 9. We might be working with numbers in the millions or trillions but the base-10, place-value numbering system allows us to break down complex problems into simple steps. Just as anyone who can write a sentence can eventually write a book if they write enough sentences, anyone who can add any two numbers from 0 – 9 can add any two large numbers with a process of easy steps. If our numbering system had a unique number for each quantity, it would have been impossible to land a man on the moon. Computers could be programmed to use this other numbering system effectively, but not humans.

Based on this analysis, it should be clear that language and mathematics are similar in important ways. Mathematics has nouns (numbers) and verbs (arithmetic operations) and it uses the equivalent of an alphabet (0 - 9) to construct numbers that continue ad infinitum. That is, rather than think of mathematics as the science of quantity, we could also think of it as the language of quantity. This should help us demystify what mathematics is and how it relates to language.

Universals

Many people are not familiar with the philosophical term universal, but most people are familiar with the content of a dictionary, which consists predominantly of universals. A universal is not another way of saying "word," because many words are not universals, but most words are universals and the power of language derives primarily from the power of universals. The problem of universals is an important issue in the history of philosophy, which most philosophers have addressed in one way or another. I have addressed the issue elsewhere, so there is no need to provide a full overview of universals, but I will address why numbers are not universals and why this is important.

Universals are about the properties of things in the world that apply to a plurality of things, like blue or rational. Universals are not proper nouns, like your name, which apply to one individual. No one disagrees that people talk about properties as a matter of fact, but the problem of universals attempts to address whether these properties have a foundation in reality (a school of thought known as realism), are merely constructs of the human mind to assist with communication (a school of thought known as nominalism), or some other possibility. For example, if we consider blue, does blue exist independent of the human mind? To answer this, we should consider the old conundrum of the tree falling in the forest and no one being there to hear it. Does the tree make a sound? It depends on what you mean by sound. If by sound we mean the human experience of sound, then no. However, if by sound we mean the sound waves resulting from the falling tree, then yes. If we had placed a microphone in the forest, it would record the falling tree. The same goes for color. If we ask whether the color blue exists independent of the human mind, we should clarify what we mean. If by color we mean the human experience of blue, then no. If by color we mean the wave frequency of blue (450–495 nanometers), then yes.

Many open-minded nominalists could probably be convinced that sound waves and color frequencies exist independent of the human mind and can be measured with scientific equipment, and are therefore valid universals. However, there would be less agreement for a more abstract term like "cat." No one disagrees that cat is a useful term to helps us understand the world of animals and to communicate with each other, but there is no objective wave or frequency for cat that we can measure with scientific equipment. The argument could be made that cat is a gap word to help us organize our thoughts, such as showing how different feline species are related to each other, to help us convert reality into convenient categories. We could consider DNA or physiological features, but where do we draw the line without ambiguity? Even if we could find an objective foundation, the decision will still be somewhat arbitrary because we could use another model to change the way we categorize living organisms.

For example, in zoology we use the kingdom, phylum, class, order, family, genus, species model to categorize living organisms, but there is nothing sacrosanct about this model. There might be a way to prove that this is the most efficient and effective model, but it applies as much to the needs of the human mind as it does to the facts of reality. This situation is easier to understand with Aristotle's ideas of substance or essence, because we can distinguish different species of cats, but the problems multiply with Plato. If we have the Forms or Ideas of Persian cats, Cats, and Animals, does this mean Persian cats participate in all three Forms or Ideas at the same time? If we breed two different breeds of cats to create a new breed, is this new breed made possible by the Form or Idea of this potential breed that has been lying dormant in the Platonic realm? Or, does the act of creating a new breed create a new Form of Idea in the Platonic realm?

The situation becomes more difficult when we ascend to higher levels of abstraction for words like "justice" or "good." We cannot measure the waves or frequencies of these abstract words and we do not have the benefit of a zoology model for them, so we have to look for different ways to understand these words. On the one hand, if we take the model of Aristotle, he would argue that words like justice or good refer to the human experience for individuals with a refined capacity for reason. In other words, these words have an organic and psychological foundation that exists potentially in our lives. If we are not just or not good it means we are not living in accordance with our rational nature. Just as we need to consume the right nutrients to optimize our biological growth and maturity, there are certain mindsets and actions that are required to optimize our biological growth and maturity. On the other hand, if we take the Platonic or religious model, then some of these words have an otherworldly foundation.

The goal of this analysis so far is to frame the issue, not to take sides one way or the other (although some degree of realism seems unavoidable), but we can now shift our attention to numbers. Given what we have addressed so far, does it make sense to say numbers are universals? If we recall that universals are about the properties of objects or the world, the answer is no because multiple numbers can apply to objects at different times. We would never point to a husband and wife and say they now possess the property of 1 and previously possessed the property of 2, and the essence of a dozen cookies would not alter if someone ate one and reduced the count to 11. To consider it from another perspective, would a dictionary benefit from having an entry for each number that we would look up if we were confused about the meaning? If so, how many numbers would we include in the dictionary?

One number is different from another number, but the difference is not essential, which is why numbers do not have a legitimate place in the dictionary. If we have two piles of gold coins, one with 9 and one with 11, nothing will change if we move one coin so that both piles now have 10 coins, whereas the situation would change if we introduced silver or bronze coins into the mix. We could argue that some numbers have properties, such as being even, odd, or prime, but this does not change the fact that a number is an abstract symbol devoid of properties. A number can represent objects with properties, but the number does not have properties, and is therefore not a universal. The one thing we can say with certainty, which Aristotle stated more than 2,000 years ago, is that the world consists of a plurality of objects—the category of quantity. That is, even though Plato's theory of Forms or Ideas is open to debate, Plato was correct to not place numbers in the realm of Forms and Ideas because numbers are not universals. We do not need a 3D printer to produce the numbers but we do need a 3D printer to create humans capable of solving mathematical problems with numbers.

At first glance, accepting the apparently obvious conclusion that numbers are not universals seems like much ado about nothing, but this is bad news for the Cantor camp. If numbers are not universals and lack an essence (or independent existence), we can then say with confidence that they are a product of the human mind, and that whatever peculiar properties numbers might have has nothing to do with the world or whatever made the world possible. There is no need to have numbers in the Platonic realm to create the world, if such a realm even exists; all we need in the Platonic realm is a mechanism to create a world with a plurality of objects and human minds that are capable of thinking quantitatively. The Cantor camp generally recognizes that a bottom-up or constructive approach to mathematics, such as beginning with 1 and +, will never take us all the way to Cantor's paradise of actual infinity. The only way to make a top-down approach to mathematics that can make actual infinity possible is to provide mathematics an abstract foundation, like set theory, but this seems impossible if numbers are not universals.

The Gaps

Just as the previous section helped us understand why numbers are not universals, we can also consider what I call the gaps to gain a better understanding of numbers. The basic idea is that as we organize our thoughts and ideas, one observation or one piece of analysis at a time, we can identify gaps that we can plug to reach meaningful conclusions that often could not have been reached otherwise.

For example, as I will address in more detail later, if we consider color, it makes sense to say we can imagine what a color in the gap looks like, even if we have never seen it before. Unless we work in a paint store and spend our days looking at color wheels, most of us will never see every possible color. (Using the RBG 0–255 model on a computer, there are 16,777,216 possible colors. We could further divide the color shades, but we could not see the difference due to the limitations of our eyes.) However, if someone shows us an array of shades of blue from light to dark and one of the colors in the middle of the array is covered with tape, we could use our imagination to approximate what the color looks like, even if we had never seen the particular shade of blue before. This raises the question of what raw material our imagination uses to construct a color we have never seen. The important point is that we can use our understanding of colors to make a reasonable guess. Then again, this might be an illusion because there is no way to verify that the color in our imagination is an exact replica of the color on the array.

If we consider astronomy, we can study planets and the way they rotate and orbit within a particular galaxy or planetary system to make conclusions about things we cannot see. For example, if we use the mathematics of Kepler's ellipses, we can make accurate calculations about how long it will take for a planet to complete one orbit. However, if we enter all the relevant data of the visible planets into a computer model, to include how the gravitational force from each planet affects the other planets, and yet we still have errors, then we can safely conclude that there must be another planet in the system that has not been observed or factored in. Not only that, we could probably estimate the precise mass and location of the mystery planet if our calculations are accurate enough, such that if we point our telescope in this location, we would probably see a planet. In other words, if we were to account for the mass and location of the new planet, then all the other errors in our calculations should go away.

If we consider a scenario of two armies fighting, then if the superior side experiences a series of surprising defeats, it could assess the various possibilities. The first possibility is that the other side was better. However, if there is no reason to reach this conclusion and the defeats do not make sense in the broader context of the war, then the army could look within for possible explanations. For example, if the army considers the three major defeats during the past year and notices that the only thing they had in common was communication with a particular logistics center for supplies and ammunition before the battle, then it might be possible that the enemy is intercepting those communications to learn about the battle plans. One way to confirm this theory would be to use a new encrypted channel for the next communication with the same logistics center and see what happens.

The army might never know for sure, but some decisions have to be made with incomplete information. This scenario is not as scientifically certain as the previous example with planets, because human behavior is involved, but the same gap analysis applies.

As we transition to language, the situation changes. Rather than consider planets or battle history, we consider language itself, such as how we sometimes impose language onto reality. Consider politics, where it is common to speak about a political spectrum from left to right. The language we use is not completely objective, such that aliens might be confused by it, but the language is meaningful. Consider the following spectrum.

Left-wing	Moderate-left	Centrist	Moderate-right	Right-wing

Most people use these terms, often without precision, but if we consider all the relevant variables along the political spectrum, then we can make these terms more precise and functional for understanding our political system. On the flip side, people will tend to resonate with one part of the political spectrum and even adjust some of their beliefs to better fit. For example, people who consider themselves Democrats or Republicans probably feel the need to assume the view of the party on issues like abortion or guns. There is no limit to the scenarios where we can use language to create meaningful distinctions that do not necessarily have an objective reference in the world. Consider rating systems that are used to evaluate employees.

Failure	Limited Success	Meets Standards	Very Good	Outstanding

I am not suggesting that words like failure and outstanding are completely subjective, but the mind seems to have a natural propensity to find a middle ground between two extreme positions and to fill the gap as necessary to achieve the right level of meaningful distinctions and precision. For the above example, we could fill in the four gaps with appropriate terms, such as "excellent" between "very good" and "outstanding," which might be useful or even necessary in some scenarios, but the spectrums we establish often reflect our own minds more than they reflect reality.

Finally, as we transition to mathematics, we gain a better understanding of numbers, in particular, what they are not. Numbers lack essence and are not universals. They are purely symbolic and can apply to different things in different situations. We would not analyze the difference between 4 and 17 the same way we would analyze the difference between Beagle and Labrador. Numbers have no properties that allow for gap analysis because whereas each word in a language must be generated and defined on its own terms, the numbers themselves can generate all the other numbers. In fact, as I will address later, every number that can be written down or used in a mathematical problem can be reduced to 1 and +. We can use the numbers themselves to fill the gaps, such as inserting 7 between 6 and 8 via addition or subtraction, but there is nothing essential about 6 or 8 that make this possible. Proposing 7 between 6 and 8 is not like understanding that "very good" is between "meets standards" and "outstanding." For example, suppose someone showed us a list of sequential numbers (like the shades of blue) and one of the numbers was covered with tape. Is there any doubt that we could say which number is under the tape? We would not require any systematic observations or power of imagination to fill the gap between 6 and 8.

Classical & Modern Logic

One of the important conclusions of this book and in the argument against actual infinity is that mathematics is not logic, more specifically, that mathematics cannot be reduced to logic. Therefore, we should first understand what logic is before we pursue this line of thinking. At its core, logic is the rules of thought, that is, laws that are true whether or not we understand them or accept them. The practice of logic involves syllogisms or other symbols to make arguments, but those are merely an exterior representation of what is happening below the surface, a visual demonstration that often serves the purpose of allowing us to extend beyond the limits of our memory or immediate grasp, for the same reason we need the painful process of long division. No matter how simple or complex our system of logic is, it should act in accordance with the laws of thought. Just as 2 objects in a room are 2 whether or not we write it down, such as 1 + 1 = 2, logic is true whether or not we write it down, but our finite minds sometimes need the process or methodology to solve the problem.

Different philosophers have identified different rules of thought, but we will focus on three classical laws to explain the general idea, not to make a definitive conclusion about logic. First is the law of identity, which states a = a. For example, if we were to argue that a triangle has three sides, this is true because triangles by definition have three sides, a = a. The only way to employ logic is to use terms that have been clearly defined and understood, such as triangle. If we take the position of Heraclitus or the relativists that everything is in a constant state of flux and that truth and meaning are relative, then the possibility of logic and truth goes out the door. Second is the law of contradiction, which states ~(a and ~a), that is, nothing can be and not be at the same time. For example, as Schopenhauer observed, which is important for this book and our analysis of infinity, no body is without extension. Third is the law of excluded middle, which states a or ~a, that is, that everything must be or not be. For example, consider carbon-based life form. We can attribute this or not to any object in existence, such as people are carbon-based life forms and rocks are not.

1. Law of Identity: a = a.
2. Law of Contradiction: ~(a and ~a).
3. Law of Excluded Middle: a or ~a.

We can have a reasonable debate about whether this list is complete or needs to be modified, but the important point is that these laws have to be discovered. They do not depend on our own thought processes; on the contrary, valid thought processes depend on them. To think of the laws of thought another way: what must we assume to be true in order for truth to be possible? In other words, any time we claim something is true, we are implicitly making at least one of these three assumptions, whether we know it or not. Once we have a general framework for the laws of thought, we can create systems of symbols and rules of inference to help us solve logical problems in a systematic way.

For most of history, Aristotle's syllogistic logic was the standard and played a key role in shaping the history of Western philosophy. This logic can be expressed in language with syllogisms or visually with Venn diagrams. If we consider any two things, A and B, they can be related in one of four ways:

1. All A are B, such as all cats are animals.

2. No A are B, such as no cats are plants.
3. Some A are B, such as some animals are cats.
4. Some A are not B, such as some cats are not black.

With logic, we are concerned with the symbols and the rules of inference, not the content, and the goal is always the same: establishing the relationship between A and C via the middle term B. If our logic is correct, then our statements will be true regardless of what we plug in for A, B, and C, as long as A, B, and C obey the law of identity (a = a). Aristotle's logic takes this framework and works out all the possibilities with a handy square of opposition so that logic becomes a plug-and-play methodology, to include telling us which combinations produce valid conclusions and which ones do not. For example, the following syllogism is always correct because if all the As are in the B bucket and all the Bs are in the C bucket, then all the As are in the C bucket.

1. All A are B
2. All B are C

∴ All A are C

The following syllogism might be correct but is never logically correct because all the As and Cs might be in the B bucket, but we are unable to establish the relationship between A and C.

1. All A are B
2. All C are B

∴ ?

We can use mathematics to show that we can construct 256 possible syllogisms, so there is no need to review them all here. The important point is that of the 256 possible syllogisms, only 24 of them are valid, less than ten percent. The power of this system is that we can filter our arguments through the lens of the 24 valid syllogisms, which allows us to dismiss arguments outright that do not fit the mold. For example, if we begin a syllogism with two "No" or two "Some" premises, then no logically necessary conclusions can follow—never. Not only this, we can use these 24 valid syllogisms to shape the way we think so that we avoid slipping into common errors and stay on the path to constructing valid arguments. Aristotle's logic is still with us today, almost 2,500 years later, largely unmodified, and is still a staple of any Logic 101 textbook.

Modern logic did not enter the stage until the late nineteenth century, but the change was less revolutionary than some imagined. The symbols and rules were modified to expand the realm of logic beyond syllogisms, but the same laws of thought still applied and logic still could not extend to things not under the rubric of logic, such as mathematics. No one attempted to modify the law of identity or the law of contradiction, but some did reconsider the law of excluded middle, which is beyond the scope of this book but has important implications for mathematics and infinity.

Rather than try to fit reality into the form of a syllogism, modern logic considers other models, such as propositional logic, predicate logic, modal logic, and mathematical logic. In the case of propositional logic, rather than focus on terms like "all", "no," or "some," we focus on terms like "and," "or," "not," or "if/then." For example, to conduct a search on the Internet, if you say "A and B" it means both A and B, whereas if you say "A or B" it means either A or B. It might mean both A and B, but both A and B are not required for the search results. "A or B" will surface the "A and B" search result but not the other way.

One of the most important rules of inference for propositional logic is Modus Ponens, which states that if A implies or causes B (symbolized as A → B) and we know that if A is true, then B must be true as well.

1. If it rains, the streets will be wet (A → B).
2. It rained (A).

∴ The streets are wet (B).

The opposite and equally useful rule of inference is Modus Tollens.

1. If it rains, the streets will be wet (A → B).
2. The streets are not wet (-B).

∴ It did not rain (-A).

It should be commonsense to conclude that it did not rain if the streets are not wet, which is how many of us check whether it rained when looking outside in the morning, but this simple formula often gets misused. The most common error is called affirming the consequent, which is the B proposition.

1. If it rains, the streets will be wet (A → B).
2. The streets are wet (B).

∴ It rained (A).

This conclusion might be true, but, and this is the key to logic, it is not necessarily true. After all, someone might have opened a fire hydrant or washed a car. This point is especially important in the correlation versus causation debate, which is prominent for many important issues that shape social and political debates. For example, suppose someone argues in favor of the death penalty because it results in fewer homicides, and then points to states with the death penalty and low homicide rates to justify the claim. The claim might be true, but it is not logically true because variables other than the death penalty might account for the low homicide rates. Therefore, we will have to look for other arguments or datasets to identify the causal variables. This might bring us back to the death penalty after all, such as states that eliminate the death penalty and see an unusual rise in the number of homicides.

Just as we can use the core components of syllogisms to construct a list of all the possible syllogisms, we can use the core components of propositional logic to construct a list of all the possible propositions. And given that logic rests on the laws of thought, we can consider ways to translate between the different systems of logic, just as we can translate between different languages. For example, if we consider streets, logic dictates that they are either wet or dry. Therefore, wet streets are a subset of streets, just as cats are a subset of animals. If we consider wet streets, they are caused by rain, open fire hydrants, washed cars, or other possibilities. Therefore, streets covered with rain are a subset of wet streets and wet streets are a subset of streets, just as Persian cats are a subset of cats. Therefore, just as we can conclude that all Persian cats are animals, we can conclude that all streets covered with rain are streets. The language can sound forced or stilted, but we have a model for syllogistic thinking within propositional logic.

1. All streets covered with rain are wet streets.
2. All wet streets are streets.

∴ All streets covered with rain are streets.

If we understand what this syllogism really says, regardless of how odd it sounds, we can conclude that rain will cause the streets to be wet and that a lack of wetness on the streets implies a lack of rain. The primary difference is that whereas syllogisms are concerned with the predicates of individual things, propositional logic is more concerned with causality (hence the rationalist attraction to logic), but there is no reason to conclude that the two forms of logic are mutually exclusive or unrelated. Depending on the situation, we can use different types of logic to help us solve different types of problems.

The Logic of Geometry & Arithmetic

When we shift to mathematics, the most important thing to remember is that one form of mathematical logic is as old as logic itself, in particular, the method of solving proofs for Euclidean geometry. A better term might be deduction, not logic, but the goal is similar: using axioms to reach necessary conclusions. Most people have had the pleasure (or pain) of solving proofs in geometry class, such as proving the equality of angles that result from a line intersecting two parallel lines. As is the case with any deductive system or logic, many of the problems are simple and can be solved by observation, but other problems are less obvious and require a deductive methodology of several steps to reach a conclusion that is not obvious from the beginning. We cannot see what is on a park bench 100 yards away, but if we take purposeful steps and get close enough to the park bench, we can say what is on it.

There is one important difference between the logic we addressed in the previous section and the logic of geometry. In the case of logic, we often begin with simple universals, such as Persian cats, cats, and animals, and then make statements like, "All Persian cats are cats" and "All cats are animals" to conclude that "All Persian cats are animals." This might not seem like a major intellectual accomplishment, but these syllogisms can extend for dozens of layers, which means we can use them to reach conclusions that are beyond our immediate intellectual grasp.

1. All A are B
2. All B are C
3. All C are D

...

25. All Y are Z

∴ All A are Z

The identification and definition of universals, such as A, B, C, and so on requires the power of the intellect, but no one would say that any of these universals have a truth value (true or false), aside from claiming whether they exist, and existence is not a predicate. Persian cats exist; it is not a truth claim about how they relate to other objects. When we shift from individual universals to the premises, we are making a truth claim about the world, such as "All cats are animals." This premise will be true for all future syllogisms and could be set aside for future use. The problem is the list of true premises continues ad infinitum, and thus trying to keep a database of all true premises would probably be a waste of time, especially if we keep in mind that we would have to combine true premises with other true premises to generate other truths, which would also continue ad infinitum.

The difference with geometry is that we begin with a short list of axioms that we can use to prove everything else in geometry. Just as we use a finite alphabet (26 letters) to generate every possible word and we use a finite set of numbers (0–9) to generate every number we use, we can use a short list of axioms to prove every truth in geometry. With the axioms we can prove everything in geometry, and every true statement in geometry can be reduced to one or more axioms.

There are two important things to remember about axioms: first, they make true statements about geometry; and second, they cannot be proved (read that again). Whereas a premise, such as "All cats are animals," makes a truth claim about the world that can be falsified, it requires a discursive thought process to arrive at this premise. We cannot simply open our eyes and see that it is true. In the case of the axioms of geometry, however, this is precisely what happens—we can open our eyes and see that they are true. For example, the claim that parallel lines never cross is self-evident. I know this claim is true and I can use this claim to prove other geometric truths, but there is no way for me to prove that the claim is true, and there is no way to collect evidence to support the claim. Parallel lines never cross—accept it. We do not require dozens or hundreds of observations of parallel lines to reach the right conclusion.

The idea of axioms raises an important point. If, as many people claim, all knowledge is grounded in sense perception, then how can we be so sure about the axioms of geometry? After all, we cannot perceive the truth of the axioms. Saying that parallel lines never cross (intellect) is not the same as saying the sky is blue (vision). As I addressed earlier in the section on Schopenhauer, one possible explanation is intuition. By intuition we do not mean hunches or mystical insight; rather, we mean the ability of the intellect to reach valid conclusions about the world that cannot be reduced to perception. Despite the similarities between logic and geometry, the most important point is that geometric proofs have nothing to do with numbers or infinity. We might use geometry to prove that one angle of a triangle is 60 degrees, but we do not use numbers or the operations of addition, subtraction, multiplication, or division to solve geometric proofs, mostly because there is nothing inherently quantitative about the continuums of geometry. Thus, in order to address infinity, we will have to turn to numbers and the logic of arithmetic.

One of the most important claims of this book, which has been confirmed by logic, is that arithmetic cannot be reduced to logic with a finite list of axioms and therefore cannot provide a foundation for mathematics. Frege argued that arithmetic could be reduced to logic by defining all the key terms, such as number and the operations, and then transferring them into a framework of logic, but Russell rained on his parade by showing the inherent contradiction in his definitions. Others like Cantor established set theory to provide a foundation for mathematics, but set theory is incomplete (there are truths in mathematics that set theory cannot prove or disprove) and therefore cannot provide a foundation for mathematics. My sense from doing research is that mathematicians were eager to establish a cohesive theory of and foundation for mathematics, just as the discipline of psychology tried to integrate some unfortunate ideas, like behaviorism, to give psychology the respectability of an empirical science. If thinking about mathematics in terms of logic or set theory helps us understand or solve problems, then more power to those who try, but this does not mean that logic or set theory provides a foundation for mathematics.

To understand why arithmetic cannot be reduced to logic, we should take a closer look at what purpose logic serves. As Russell and others were eager to show, everyday language has problems baked into the cake that result in confusion and ambiguity. (Russell was fond of paradoxes that told us less about the world and more about the self-referential challenges of language, such as the set of all sets that are not members of themselves or the barber who shaves all and only those who do not shave themselves.) The purpose of logic is to strip away all the confusion and ambiguity of everyday language, resulting from the bumpy road of history and the quirks of the human mind, to develop a pristine symbolic language. Regardless of the type of logic (syllogistic, predicate, modal, etc.), a functional logical system will clean up language and allow us to reach clear conclusions.

Once we establish a functional logical system, it would be absurd to ask about the logic of the logical system. Would anyone propose studying the logic of syllogisms or the logic of geometric proofs? The same goes for arithmetic; it is already a pristine symbolic language of quantity that is (mostly) devoid of confusion or ambiguity. Arithmetic was not derived from an everyday language of quantity that was full of confusion and ambiguity. I would not equate arithmetic with logic, but the bottom line is that arithmetic is already a pristine language, which means it is not in need of a reduction to logic.

A second reason to understand why arithmetic cannot be reduced to logic is that the core elements of arithmetic, 1 and +, can generate the rest of arithmetic. To look at it another way, all of arithmetic can be reduced to 1 and +. Logic usually involves identifying unique truth nodes in the world and linking them into a complex web or hierarchy of truths, via syllogisms or rules of inference, such as concluding that cats and alligators are animals. That is, with logic first we have to observe and map out reality and then use the methodology of logic to link all the parts. We cannot begin with the word cat and use it to generate every other word in the English language. With arithmetic, however, we can do just that. If logic wants to find a link between the concepts "Persian cat" and "animal," we have to generate the intermediary concept "cat" to fill the gap. With arithmetic, however, if we want to fill the gap of 6 and 8, we can do so via $6 + 1 = 7$ or $8 - 1 = 7$, which requires no new understanding or analysis of reality because numbers do not have an essence. The truths of arithmetic are generated from within arithmetic.

A third reason to understand why arithmetic cannot be reduced to logic is there is no need to define the terms in the same way we would for other disciplines. Despite the noble efforts of Frege, Russell, and others, there is no reason to define number with the same rigor as we do for universals. The majority of people have no idea about the historical attempts to define numbers and yet they can solve mathematical problems with ease. The reason for this is that we understand arithmetic intuitively; it is not a discursive thought process. No one is confused about the meaning of 1 + 1 = 2, and no definition will assist. Just as we know that parallel lines never cross, we know what numbers mean. If we look at a six-sided die with dots, we understand what we mean by 1, 2, 3, 4, 5, and 6. We use numbers to represent numbers we cannot grasp, such as 24,538,912, but we can use this number in any problem without missing a beat. On the other hand, we need precise definitions for every element of logic to perform logical proofs. Likewise, we do not need a clear definition for add, subtract, multiply, or divide. If a child grabs one cookie from a jar, eats it, and grabs one more, he performed addition in his stomach and subtraction from the cookie jar. Addition and subtraction are inverse activities, something we do.

My sense is that the attempt to reduce mathematics to logic is motivated by a desire to provide a top-down foundation for mathematics rather than a bottom-up foundation, such as that provided by Aristotle's category of quantity. The reason for this is that actual infinity is impossible with the bottom-up, potential infinity mathematics of Aristotle, as the Cantor camp will admit. If we begin with 1 and +, we will never arrive at actual infinity and Aristotle's idea of potential infinity will be proven correct. To make actual infinity possible, we need a top-down foundation for mathematics, which necessarily takes us to the realm of Platonic Forms or Ideas, and one way to achieve this is to provide a logical foundation for mathematics. As I have argued and will argue again in the last chapter, there is no need to postulate the existence of an actual infinite set of numbers in a Platonic realm to account for the idea of infinity in mathematics.

Conclusion

Mathematics is the pristine language of quantity. Regardless of language or culture, 1 + 1 = 2 is a clear and unambiguous truth that will not benefit from definitions or the framework of set theory. This does not mean mathematics cannot produce any paradoxes or peculiar results, such as the idea of actual infinity or the claim that .999... = 1, which I will address in more detail later. Although the foundation of mathematics—the plurality of objects in the world (the category of quantity)—is independent of the human mind, the language of mathematics (base-10, place-value numbers) is a product of the human mind. The same goes for language. Things like cats and rivers exist independent of the human mind and provide the foundation for language, but language is a product of the human mind. Mathematics and language serve the two-fold purpose of providing a mechanism to construct an abstract model of the world and provide a modular foundation (letters and numbers) that allow us to map reality ad infinitum and transcend our own epistemological limitations. Our minds cannot distinguish a 1,000-sided polygon (a chiliagon) from a 999-sided polygon, but mathematics allows us to distinguish them with precision.

Chapter 4: The Numbers

Most people absorb the complex vocabulary and grammar of their native language effortlessly, often without awareness. They can correctly use the past perfect tense or the subjunctive mood but might not be able to explain what these things mean. Why do we say, "If I were..."? This is why learning a foreign language is so invaluable: it forces us to think about our native language and why we do the things we do, which in turn helps us use our native language more effectively and recognize errors more easily. The same goes for numbers. Most of us from birth are immersed in base-10, place-value numbers and take it for granted the rest of our lives because it is the universal language of mathematics. Most people understand that computers use a binary numbering system (0 and 1), but most of us are unable to solve even the most basic problems in binary math. There is really nothing to be gained by tweaking our minds to think in terms of other numbering system. In theory, we could switch to a base-12 numbering system, but we would not benefit from this in the same way that we would benefit from learning a foreign language.

When analyzing something, like numbers, we have to consider the foundation of what we are analyzing and how we get to know it as a matter of fact. For example, we know that the earliest records of numbers are the natural numbers—1, 2, 3, and so on. Over time, we can observe the numbering system expand into integers and rational numbers, or even real or complex numbers. Likewise, mathematics begins with simple operations like addition, subtraction, multiplication, and division before expanding into exponents, trigonometry, derivatives, and integrals. This would suggest that simple mathematics provides the foundation for complex mathematics. This is not always the case, however, because sometimes the origins are false and get fixed over time. Consider the earliest theories about chemistry and physics, which have since been discredited.

The difference with mathematics was we never really corrected the earliest theories—never had to. Adding, subtracting, multiplying, and dividing numbers still works just as well as it did for people in the ancient world. However, just as the earliest people used vocabulary and grammar they did not understand in the same way that modern linguists do, the earliest people used mathematics they did not understand in the same way we do today. For example, the ancient Greeks were mixing and matching natural numbers with real numbers and the infinity of addition (arithmetic) with the infinity of division (geometry), often with a sense that numbers possessed a mystical power. The process of analysis allows us to disentangle and demystify these variables and put them into a hierarchical framework to better understand their origins and foundations.

Russell noted that there were two ways to approach mathematics. "The more familiar direction is constructive, towards gradually increasing complexity: from integers to fractions, real numbers, complex numbers; from addition and multiplication to differentiation and integration, and on to higher mathematics. The other direction, which is less familiar, proceeds, by analyzing, to greater and greater abstractness and logical simplicity; instead of asking what can be defined and deduced from what is assumed to begin with, we ask instead what more general ideas and principles can be found, in terms of which what was our starting point can be defined or deduced."[21] We are still in the process of analyzing mathematics, but our analysis will show that the constructive, potential infinity methodology of Aristotle was correct. When we climb to higher levels of abstraction and metaphysics—yes, we have no numbers.

Visualizing the Numbers

The argument for actual infinity hinges on the claim that an infinite set of numbers exists (in a Platonic realm). Actual infinity does not exist in the

[21] Russell, Bertrand, *Introduction to Mathematical Philosophy*, 6.

material world because the total number of atoms in the universe (some estimates are 10^{78} to 10^{82}) is a trivially tiny number in the context of infinity, where we can imagine numbers with trillions of digits. And as I will argue in this chapter, an infinite set of points cannot be evenly distributed along a finite continuum.

By analogy, my guess is that most people would accept that the letters "c", "a," and "t" or the word "cat" do not exist independent of the human mind. Everyone should agree that cats exist independent of the human mind, or that the Form or Idea of Cat exists in the Platonic realm, but the word cat exists only in our minds. The same goes for numbers, but many people (Platonists, for example) see nothing unusual about believing that numbers exist independent of the human mind even though our base-10, place-value numerical system is clearly the product of the human mind and reflects the strengths and weaknesses of how we process quantitative information. Granted, the world exhibits quantitative regularity or patterns that suggest we live in a mathematical world, but this is not the same thing as saying that numbers exist independent of the human mind.

To understand why this this is so, we should take a closer look at what it means to say numbers exist. We can all agree that cats exist in the world and that these cats provide the foundation for our use of the word cat, not the other way. The question: is there anything similar to cats in the world that provide a foundation for numbers? The simple answer is the plurality of objects in the world (the category of quantity). If 2 apples fell from a tree yesterday and 3 apples fall from a tree today, 5 apples have fallen from the tree, regardless of whether any humans exist to count them.

Yesterday: * *
Today: * * *

Granted, when I say 5, I mean (* * * * *), not "5." The symbol 5 is a human creation, just as the word "cat" is a human creation. In fact, I would even go as far as saying that we could do mathematics (albeit to a more limited degree) without using numbers. Consider dice games like craps. Most people can glance at the holes drilled into each side and add the two quantities in the blink of an eye, which means that with some training we could probably learn to process larger quantities without symbolic numbers.

The Cantor camp might argue that we cannot simply dismiss the power of numbers to shape the world we see around us, such as prime numbers or the Fibonacci sequence. As we study nature, patterns emerge suggesting that numbers must somehow account for the world around us. For example, regarding the apparent power of prime numbers, we should consider what it means to be a prime number. We know that a prime number can be divided only by itself and 1, but what does this really mean for mathematics and the world? Consider the number 12.

1 x 12 ************

2 x 6 ******
******

3 x 4 ****
****
****

It is no accident that our clocks use the numbers 12, 24, and 60: they are easily divisible, which makes it easy to think in terms of simple fractions. In the case of prime numbers, the opposite is the case.

1 x 13 *************

If we try to make two rows (2 x 6 remainder 1) or three rows (3 x 4 remainder 1), we will never get an equal number in each row.

If we consider the 13-year and 17-year periodical cicadas, we see an example of prime numbers in nature. There is some debate about why the life cycles are prime numbers, but the cicada populations supposedly survive better when they do not mix with other cicada populations, and one of the best ways to do this is to have life cycles that are prime numbers. This way, the 13-year and 17-year populations surface at the same time every 221 years (13 x 17).

There is no reason to suspect that prime numbers somehow magically shape the world. On the contrary, the safer bet was that trial and error played a key role and that the cicada populations that happened to have the most off-cycle breeding years survived the longest. That is, we can see patterns of prime numbers in the world without claiming that prime numbers exist in a Platonic realm or are actively shaping the world. Likewise, although every natural number can be expressed as the product of prime numbers (24 = 2 x 2 x 2 x 3), this does not mean that prime numbers provide a foundation for the natural numbers (24 is not composed of three 2s and one 3). On the contrary, the prime numbers are a subset of the natural numbers that just happen to have this property, and some of the peculiar properties of numbers are a function of the fact that we use a base-10, place-value numbering system. When we dig our way down to the raw foundation of quantity, a lot of the mystery disappears.

Shifting to the Fibonacci sequence (1, 1, 2, 3, 5, 8, 13, 21, ...), we see the same thing. It is true that nature often displays the Fibonacci sequence, but this does not mean the Fibonacci sequence of numbers exists independent of the human mind in a Platonic realm or actively shapes the world. First, consider a trail in the woods. As random people walk through the woods, people see the footsteps of others and use the same steps, often to avoid snow, water, or mud. At some point, the appearance of a trail steers more people to reinforce the trail, even if we learn after the fact that the nascent trail is not the optimal trail. In other words, previous activity has a cumulative impact.

Second, consider a winding river. As it flows, it makes contact with rocks, shores, and bends, which in turn shape the way the river flows. I cannot speak for Mother Nature, but the same idea applies. There are many things happening in nature, but some things get traction and gain momentum. Just as a trail gains traction when one person follows the steps of a previous person (1 + 1), the appearance of two sets of footsteps might attract two more sets of footsteps (1 + 1 + 2), which in turn could attract three more sets (1 + 1 + 2 + 3), and so on. The next five people (2 + 3) are attracted to the trail by the two most recent sets of footsteps and can no longer distinguish the first two sets (1 + 1), which suggests that Mother Nature might weigh the two most recent quantities with a greater weight.

I admit this is somewhat speculative and do not consider it an integral part of my theory, but it serves the purpose of showing that the plurality of objects in the world can shape events due to the properties of quantities, without invoking immaterial numbers in a Platonic realm. If we consider life on earth, think about how much it is influenced by the mathematics of the earth's rotation and orbit around the sun. There is nothing magical about 365 days; it just happens to be how many times the earth rotates while completing one orbit around the sun. The number 365 does not shape the process from a Platonic realm. Just as cats exist independent of the human mind and do not require the Form or Idea of Cat in a Platonic realm for their existence, quantities of objects exist independent of the human mind and do not require numbers in a Platonic realm for their existence.

From Natural to Rational

The earliest people were presented with a plurality of objects. Just as it took a long time for people to discern patterns in reality to give names to things, it took a long time to think systematically about the world in terms of quantity. One of the biggest challenges was thinking about different objects as one group that can be counted. For example, if we look outside and see a tree, a bird, a horse, and a cloud, today we might think 4, but earlier people might find that surprising—counting dissimilar objects. Even Count Dracula on Sesame Street counted only similar objects. In fact, our language still has vestiges of this way of thinking. We still speak about a den of lions, a gander of geese, or a blessing of unicorns, suggesting that the idea of counting applies only to similar things or objects. The important point is that whereas some might suggest that the natural numbers are a simple or unrefined level to begin mathematics, I would argue that the natural numbers changed the world and are a marvel of abstract thought.

Before starting with the natural numbers (1, 2, 3, ...), we should take some time to consider the basic mathematical operations—addition, multiplication, subtraction, and division—because the numbers and the operations go hand in hand. If numbers are the nouns of the mathematical sentence, the mathematical operations are the verbs. The interesting part is the four basic operations are all forms of addition: subtraction is the inverse of addition; multiplication is a special form of repeated addition; and division is the inverse of multiplication. As a result, just as all the natural numbers can be reduced to 1, all the basic mathematical operations can be reduced to addition.

As long as we limit our mathematical operations to addition and multiplication, there is no need to expand beyond the natural numbers. If we add or multiply any two natural numbers, the result will always be another natural number. When we introduce subtraction, however, we have to expand the natural numbers to allow for the possibility of negative numbers. For example, $3 - 5 = -2$. At first glance, a negative number might seem odd or without meaning, but we use negative numbers all the time in our day-to-day living, but in an abstract way. For example, suppose someone asks you how much money you have. If you have $1,000 in your checking account and a $2,000 balance on your credit card statement, you could say your net assets are -$1,000, such that you would have to add $1,000 to your checking account to get back to $0. You cannot touch -$1,000, but it represents a meaningful idea in the world of numbers and might generate some real pain in your life.

When we first start using numbers, we do not presume a continuous number line from 1 to ad infinitum, although this is how it is often taught in school. Over the years we might use 1, 2, 4, 7, 25, 100, and hundreds of other numbers, but we are able to eventually generalize to draw a number line to fill in the gaps.

1 2 3 4 5 6 7 8 9 10...

These are the natural numbers, symbolized by "N." They continue in one direction ad infinitum without presuming the existence of actual infinity. In fact, the idea of actual infinity seems to be ruled out: by definition, for any natural number n, as n goes to infinity, $n + 1 \neq n$. We can always add 1 to a natural number, which will always result in the number growing by 1. In this way, the idea of potential infinity seems to win the day, and this is precisely why the Cantor camp has to transform the way we define or understand numbers to allow for the possibility of actual infinity. As we will see, according to the Cantor camp, the natural numbers or the set of natural numbers can be thought of as a complete set, such that adding 1 to the set does not change the size of the set, such that $n + 1 = n$. Thus, making the leap from potential infinity to actual infinity is not achieved with the tools of mathematics and requires a rejection of our most basic intuition, which means we should consider the possibility that the Cantor camp is projecting the rules of finite mathematics onto the rules of potential infinity to achieve their objective.

The same goes for subtraction. Our correspondence and ledgers might include numbers like -6, -22, -100, but we can eventually generalize to fill in the gaps with a number line, which takes us to the integers, symbolized by

"Z." They continue in both directions ad infinitum without presuming the existence of actual infinity.

...-5 -4 -3 -2 -1 0 1 2 3 4 5...

The last arithmetic operation is division. In many cases, division results in integers. For example, 10 ÷ 5 = 2 and -6 ÷ 3 = -2. However, there are other times when the results force us to expand the numbers again to allow for fractions or decimals that terminate or repeat (more on this later). For example, 1 ÷ 4 = 1/4 or 0.25 and 3 ÷ 2 = 3/2 or 1.5. These numbers also make sense. If we take a whole pizza as a unit of 1, then it makes sense to eat 1/4 or 0.25 of a pizza. However, if we take 1 slice of pizza as a unit, then it makes sense to say we eat 2 slices of pizza, which means 2 slices = 1/4 or 0.25 of a pizza. Likewise, if it makes sense to say we are $1,000 in debt, or -$1,000, it also makes sense to say we are $1,000.75 in debt, or -$1,000.75. When we make the transition from integers to fractions or decimals, we reach the rational numbers, symbolized as "Q." Every rational number can be expressed as the ratio of two integers, p and q, such that p/q. However, as we will see, when we consider the rational numbers as a set of decimals rather than as a set of fractions, the rules change, especially in the context of infinity.

Given that we can list fractions or decimals ad infinitum between any two integers, it is more difficult to imagine a number line for the rational numbers.

...-2 ... -1 ... 0 ... 1 ... 2...

Normally, the process of generating the set of rational numbers between 0 and 1 proceeds by listing all the possible fractions for every natural number denominator, beginning with 2 (1/2), 3 (1/3, 2/3), 4 (1/4, 2/4, 3/4), 5 (1/5, 2/5, 3/5, 4/5), and so on. This ad infinitum process continues until n (1/n, 2/n, 3/n, ... , (n - 1)/n).

However, given that we can also think about rational numbers as decimals, we can use a similar process with denominators that are exclusively powers of 10, which makes sense because decimals presume powers of ten. For example, if we begin with a denominator of 10 we have 0.1, 0.2, 0.3, ... , 0.9 (every one digit decimal). With a denominator of 100 we have 0.01, 0.02, 0.03, ... , 0.99 (every two digit decimal). With a denominator of 1,000 we have 0.001, 0.002, 0.003, ... , 0.999 (every three digit decimal), and so on ad infinitum. The most important point is that both of these processes continue ad infinitum and no rational numbers will be missed along the way. The Cantor camp might balk at equating the set of fractions and the set of decimals as rational numbers, such as claiming that this methodology misses repeating decimals like 1/3 = .333..., but we will later use the power of mathematical proof and the Cantor camp's own assumptions to justify this claim. In short, if 1/3 = .333..., which we are told is an infinite decimal, then we can use an iterative process to generate .333... as well.

In some ways, this is the end of the road. The mathematics of rational numbers and the four arithmetic operations (additions, subtraction, multiplication, and division) is a closed system, like a chessboard with pieces. Within this closed system, rational numbers will always generate other rational numbers when we use the arithmetic operations. If we select any two rational numbers, we can add, subtract, multiply, or divide them, and the result will always be another rational number. As long as we play by these rules, we will never need to expand the numbering system again, and we can achieve whatever degree of precision is required to solve any mathematical problem we can imagine. For example, if we need 6,000,000,000,000 decimal places of accuracy to solve a problem for physics (which is never the case), the rational numbers will suffice.

Moving forward, as we make the transition from rational numbers to real numbers, there are two ways to do this. First, we can add more mathematical operations, such as exponents or trigonometric functions, which will generate irrational numbers. Second, we can impose the infinity of addition (the infinity of rational numbers) onto the infinity of division (the infinity of the continuum) to generate the set of real numbers by "filling in the gaps" that allegedly exist in the set of rational numbers. That is, although there are no mathematical operations that can generate the set of real numbers (which should raise concerns), we can supposedly assume that the real numbers can "fill in the gaps" allegedly left by the rational numbers. (As I will prove later, it does not make sense to say the set of rational numbers leaves gaps on the continuum or that the real numbers can fill the alleged gaps left by the rational numbers.)

The Real Numbers

The theory of actual infinity espoused by Cantor really comes into play with the set of real numbers (the infinity of the continuum), which includes the rational numbers and the irrational numbers. The process of imposing discrete numbers onto a smooth continuum and filling the gaps is known as "arithmetizing" the continuum, which in many ways is the most important bone of contention in this book. If you reject this idea, you probably support the idea of potential infinity (the discrete can never mysteriously transform into a continuum, regardless of how we define our terms). If you support this idea, you probably support the idea of actual infinity (the discrete can mysteriously transform into a continuum, with rigorous definitions to support it).

In short, Cantor argued that, although the natural numbers, the integers, the rational numbers, and the real numbers can all be thought of as infinite sets, there are more real numbers than the other sets of numbers (the size or cardinality of the real numbers is larger), as I will address in the next chapter. As I said in the previous section, the rational numbers and the arithmetic operations of addition, subtraction, multiplication, and division are a closed system, meaning that we can never escape from the rational numbers and there will be no need to expand the set of rational numbers if we limit our work to the four arithmetic operations. Therefore, if we are going to make the transition from rational numbers to real numbers we will have to introduce new mathematical operations. However, this alone will not suffice.

Although Cantor's theory of actual infinity focuses on the real numbers and irrational numbers, it focuses on a particular type of irrational number—transcendental numbers. Before we arrive at the transcendental numbers, however, there is a gray area of irrational numbers called algebraic numbers that Cantor would put in the same category as rational numbers in terms of being "countable" or "smaller" than the set of transcendental numbers. To make the transition from rational numbers to the algebraic numbers, we add the polynomial equations of algebra, such as $x^2 + 4 = 0$. We continue to use rational numbers and the arithmetic operations of addition, subtraction, multiplication, and division, but we add the exponent (x^n) operation, which is a form of multiplication in the sense that we multiply an unknown number times itself n times. Given that multiplication is a special case of addition, an exponent is also therefore a special case of addition, which will prove to be an important claim when we address the power set.

The use of a variable like x might seem like a leap of abstraction, but it is often just a way of working a problem backwards if we already know the answer. Normally, in an addition problem, we add two known numbers to arrive at the unknown answer, such as $5 + 7 = x$, but sometimes we know the answer and need to know how we got there. For example, if we know that two people ate a pizza with 12 slices and one person ate 7 slices, then how many slices did the other person eat? We can write this as $x + 7 = 12$ and use the tools of algebra to solve for x. Or, if we are clever, we can express the problem as $12 - 7 = x$ and skip the algebra. As long as we do this, we will stay within the chess game of the rational numbers, but the introduction of a new type of multiplication, the exponent (an unknown number multiplied itself n times), opens the door for algebraic numbers. For example, consider the following problem.

$x^2 - 2 = 0$

$x^2 = 2$ (add 2 to both sides)

$x = \pm\sqrt{2}$ (take the square root of both sides)

Therefore, we can solve this problem if x equals $\sqrt{2}$ or $-\sqrt{2}$. As it turns out, however, $\sqrt{2}$ is not a rational number because the decimal repeats forever without repeating. There is no rational number p such that $p \times p = 2$. Of interest, we can duplicate this same model for every rational number, to include $\sqrt{3}$, $\sqrt{4}$, $\sqrt{1/3}$, $\sqrt{1/4}$, and so on. Some of these have rational solutions, such as $\sqrt{4} = 2$ and $\sqrt{1/4} = 1/2$, but more often than not the results are irrational. We can also duplicate this same model for every power, to include x^3, x^4, x^5, $x^{1/3}$, $x^{1/4}$, and $x^{1/5}$. The key point is that although the methodology introduces new types of numbers, polynomial equations are in many ways a natural extension of the arithmetic operation of multiplication, which is a special form of addition. I will address the significance of this later, but this is the reason the set of algebraic numbers are "countable" and "smaller" than the set of transcendental numbers.

When we transition to transcendental numbers, we arrive in the realm where Cantor took inspiration for the most controversial claims of his theory of actual infinity. For the purposes of this book there is no need to address the intricacies and nuances of transcendental numbers (they never repeat or terminate after the decimal and are not algebraic numbers), but we know that trigonometric functions can generate the most famous transcendental number, π, and other functions can generate other transcendental numbers like e.

The first odd thing about transcendental numbers is that although the size of the set is supposedly significantly greater than the set of natural numbers, there is no method to systematically generate a list of transcendental numbers, and there are some numbers that are suspected of being transcendental but it is difficult to prove. Granted, mathematicians are honest enough to admit that it is difficult to prove whether some numbers are transcendental, but this raises questions about why Cantor and others are so confident that the set of transcendental numbers is so massively large that it absolutely dwarfs the natural numbers. As we will see in the next chapter, this has to do with calculating the size or cardinality of sets of numbers.

The second odd thing about transcendental numbers is that the same rules that apply to the rational numbers for the arithmetic operations should also apply to the real numbers, but it is not clear how to perform a simple problem like π - e because the decimals for π and e continue forever without repeating, which makes subtraction impossible. If π and e are real numbers, there should be a repeatable way to calculate their value with infinite precision, such that they are greater than or less than every other real number and can be put through the meat grinder of the arithmetic operations to generate other real numbers. In other words, by making transcendental numbers so difficult to generate, it raises questions about whether they are really numbers. Cantor claimed we have to accept the idea of actual infinity if we accept the existence of irrational numbers, so this point is key to the debate. However, if we consider that Cantor claimed an infinite decimal like an irrational number could be thought of as a complete set, such that adding one more number to the set or decimal does not change it ($n + 1 = n$), it raises questions about what it means to say an irrational number never repeats or terminates. If the set of natural numbers can be thought of as a complete set, then so can the decimal numbers of an irrational number.

This brings us full circle to the difference between the infinity of addition and the infinity of division, between quantity and the continuum. When we impose the rational numbers onto a continuum, the claim of the Cantor camp is that there are many "gaps" that can be "filled" with the real numbers to make the line "smooth." To justify this claim, the Cantor camp ignores the metaphysics of what they are doing and attempts to resolve the problem with rigorous definitions, such as least upper bounds. However, the process of imposing an infinite set of points onto a finite continuum has metaphysical consequences, as Zeno demonstrated. If the points have dimension, no matter how small, then the continuum will necessarily extend to infinity to accommodate them. On the other hand, if the points are dimensionless, they will necessarily collapse to 0 and the continuum will be empty. We could introduce infinitesimals to solve the problem, which I will address in more detail later, but this will not resolve the logical contradiction.

Understanding the Real Numbers

The sticking point in making the transition from the rational numbers to the real numbers, from the discrete to the continuous (that is, "arithmetizing" the continuum, if it even makes sense to propose such a thing), boils down to recognizing that we are really making the transition from the infinity of addition to the infinity of division, as explained by Aristotle. These two systems are metaphysically distinct. Whereas discrete objects are inherently quantitative and can be counted, there is nothing inherently quantitative about the continuum, even after we arbitrarily impose a numbering system onto it and attempt to bridge the gaps with least upper bounds. That is, rather than try to "arithmetize" the continuum by defining the real numbers in terms of the rational numbers, we could impose an ad infinitum iterative decimal process onto a continuum similar to what we did with the rational numbers, with the understanding that we are still imposing something of a different kind onto the continuum.

One way to do this is to recognize that we have chosen to express real numbers as base-10, place-value decimals (I am not aware of a better way), which means we should focus on fractions with denominators that are powers of 10, such as 1/10, 1/100, 1/1,000, and so on, to generate our decimals. If we imagine a continuum on which we have marked one point as 0 and another point as 1 (this is always arbitrary), then we can divide the continuum between 0 and 1 with decimals that are powers of 10.

[0.0 0.1 0.2 0.3 0.4 0.5 0.6 0.7 0.8 0.9 1.0]

If we focus on the continuum between 0.0 and 0.1, we can continue the iterative process.

[0.00 0.01 0.02 0.03 0.04 0.05 0.06 0.07 0.08 0.09 0.10]

This process can continue ad infinitum and can be used to generate every possible decimal between 0 and 1 one decimal place at a time; all the tenths, hundredths, thousandths, and so on. At each step, the size of the set of decimals grows by a power of 10; 10 (one decimal place), 100 (two decimal places), 1,000 (three decimal places), or 10^n after n decimal places. There is no need to define the real numbers in terms of the rational numbers or introduce ideas like least upper bounds to fill the alleged gaps. If Cantor was right that actual infinity is real and that transcendental numbers like π exist and extend to actual infinity, then this iterative process will generate π as the process extends to infinity, just as 1/3 generates 0.333... as the process extends to infinity.

I will explain this process in a different way to demonstrate why infinity really is a metaphysical problem, not a mathematical problem. Rather than consider numbers, however, we will consider points and ranges along a continuum because Cantor's theory of infinity hinges on the existence of an infinite set of points along a finite continuum. Suppose I set 10 billiard balls (ranges) separated by points on a continuum from 0 to 1, such that each billiard ball has a range of 1/10. The 10 billiard balls and the nine points between them add up to 1, which means the points are dimensionless.

Now, suppose we double the number of billiard balls that are half the size along the same continuum between 0 and 1. In this case, the 20 billiards balls and the 19 points between them add up to 1, which means the points are dimensionless.

We can continue this process ad infinitum to achieve whatever degree of precision is required for a particular problem, without presuming the existence of an infinite set. There is no theoretical limit to how far we can further subdivide this process using smaller billiard balls—1/4, 1/8, 1/16, etc. The interesting part is that as long as the ever-shrinking billiard balls continue to fill the space between 0 and 1 along the continuum (the size of the billiard balls is always offset by the quantity of the billiard balls, such that $n \times 1/n = 1$), we are still in the realm of the rational numbers. However, if we ask ourselves what happens when this process arrives at actual infinity, this can happen only when the size of the billiard balls shrinks to actual zero. However, in this case, if we add up the infinite set of dimensionless billiard balls (0) and the infinite set of dimensionless points separating them (0), the range of the billiard balls collapses to 0, which is another way of saying there is no way to evenly distribute an infinite set of points along a finite continuum.

One of the interesting findings of this study of infinity is that the methodologies we use can actually shape the results, which should not be the case if the mathematics is valid. For example, if we think about how many billiard balls we can squeeze into a finite continuum (0 to 1), it turns out that different methodologies generate different results. To show what I mean, we can start with one billiard ball that fills the continuum between 0 and 1. Next, we can fill the same continuum with two billiard balls that are half the size. Next, we can fill the same continuum with three billiard balls that are one third the size, and so on, with the following results:

1, 2, 3, 4, ... , n (n represents the quantity of billiard balls squeezed into the continuum after n iterations).

If we start with two billiard balls and double the quantity each step, we get the following results:

2, 4, 8, 16 ... , 2^n (2^n represents the quantity of billiard balls squeezed into the continuum after n iterations).

If we start with three billiard balls and triple the quantity each step, we get the following results:

3, 9, 27, 54 ... , 3^n (3^n represents the quantity of billiard balls squeezed into the continuum after n iterations).

If we start with 10 billiard balls and grow ten-fold each step, we get the following results:

10, 100, 1,000, 10,000, ... , 10^n (10^n represents the quantity of billiard balls squeezed into the continuum after n iterations).

The interesting point is that if we accept for the sake of argument that infinite sets are possible, then it would seem reasonable to conclude that each of the above four methodologies would result in the same quantity of billiard balls being squeezed into a finite continuum. After all, the first example grows at a slower rate but eventually catches up with the other three examples. However, if we allow the four examples to go to actual infinity ($\aleph$, $2^{\aleph}$, $3^{\aleph}$, $10^{\aleph}$), then according to Cantor's theory of infinity (more on this later), these four results are not the same size! In fact, Cantor would argue the following: $\aleph < 2^{\aleph} < 3^{\aleph} < 10^{\aleph}$. Thus, if different methodologies produce different results, we should have concerns about the results. This simple example demonstrates a powerful truth: either infinite sets do not exist or all infinite sets are the same size, which contradicts Cantor's theory of infinity.

Conclusion

The key take away from this chapter is the understanding that numbers (in particular, base-10, place-value numbers) are a product of the human mind and have no independent or metaphysical existence. Numbers do not exist in a Platonic realm, just as words do not exist in a Platonic realm. We create numbering systems to solve problems and better understand the world via the science of quantity. There is limited debate about the transition from natural numbers to integers or from integers to rational numbers, but there is no shortage of debate about the transition from rational numbers to real numbers, which is really what this book is about.

My goal in this chapter was to show that the transition from the discrete to the continuous is not possible in the way the Cantor camp would have us believe. We cannot "arithmetize" the continuum or define our way to infinity if the definitions have metaphysical implications that result in a logical contradiction. When we say the ratio of the circumference of a circle to its diameter (π) continues ad infinitum without terminating or repeating, this is all it means, most likely due to the incompatibility of the infinity of addition and the infinity of division. It does not mean the irrational number has an objective, complete existence independent of the iterative process that generates it. Every mathematical problem involving π will use an estimate of the number, perhaps 10 digits, perhaps 1,000 digits, which are crude approximations in the context of infinity but have a stunning degree of accuracy in the context of the universe.

Chapter 5: Static Infinity

The Cantor camp recognizes that a discrete process of counting the natural numbers (1, 2, 3, ...) never reaches actual infinity, which is why they do not agree with the constructive approach (bottom-up) or potential infinity mathematics of Aristotle. Rather than reach actual infinity via discrete counting, the Cantor camp thinks of actual infinity as a pre-existing, complete set (top-down), which leaves us with numbers most likely existing in a Platonic realm (Cantor was a Platonist for numbers), not in the material world. After all, the universe contains a finite quantity of particles, so if actual infinity exists as a complete set, it has to exist outside of the limits of the physical universe, whether in a Platonic realm or the human mind.

As is often the case, important debates often boil down to how we define our terms. If someone tells you a movie is "good," it is difficult to know exactly what this means if you do not understand what the person means by "good." Perhaps the person thinks character arcs are bourgeois and you think they are essential to the craft of storytelling. Likewise, if we want to argue that actual infinity is a meaningful term, we should first try to understand what the Cantor camp means when they talk about infinite sets.

When we attempt to discredit someone's argument, we should focus on their assumptions and definitions because if these can be discredited the argument collapses. As often happens, one side in a debate takes their own assumptions and definitions as a given and struggles with understanding the topic from another perspective, which means that one of the challenges is framing the debate from an objective perspective. The Cantor camp can talk about how Cantor's conclusions follow necessarily from his assumptions and definitions, but the conclusions are not valid if the assumptions and definitions are invalid or produce inconsistent results. For example, whereas Cantor claims he can use the power set of the natural numbers to prove that the cardinality of the real numbers is $2^{\aleph}$, I will use what I consider a better and easier method to prove it is $10^{\aleph}$.

The challenge with the debate about infinity is that the Cantor camp has managed to gain the high ground to assert its own assumptions and definitions. This allows them to frame the debate in their own terms, which makes it difficult to contradict them or for other perspectives to see the light of day. Every attempt to discredit their assumptions and definitions is countered with their own assumptions and definitions, such as claiming that actual infinity must be real because the Axiom of Infinity says it is real, or that power sets of infinite sets exist because the Axiom of Power Set says they are real.

Therefore, to challenge the assumptions and definitions of the Cantor camp, we should consider the debate from other perspectives. Just as the perspective of NASCAR drivers and the audience is shaped and limited by the fact that the cars drive counter-clockwise, the world looks different if we drive clockwise. For example, the Cantor camp represents the idea of infinity with different symbols (∞, ω, $\aleph$), but in the case of the natural numbers they do not seem interested in what infinitely large natural numbers look like, which, it turns out, is revealing.

Countable Sets

As I noted in the Introduction, Cantor's theory of infinity changes the way we think about mathematics and infinity because it transforms mathematics from a science of quantity into a science of sets, which supposedly allows for the possibility of infinite sets. In fact, as Cantor knew, set theory serves no purpose without infinite sets. Knowing the size or cardinality of a set does not help us solve applied mathematical problems, but without this definition, Cantor's structure collapses like a house of cards.

There is no doubt that thinking about the size or cardinality of sets is interesting, but this does not mean the methodology is valid or without flaws, especially when we keep in mind that set theory does not provide a foundation for mathematics. As we will see, some of the mystery and paradoxes of infinity are obscured by the symbols that are used to express it. Cantor's symbol for the smallest infinite set ($\aleph_0$, but I will use $\aleph$ in this book) looks like the generic mathematical symbol "n" and seems less scary or threatening than the traditional symbol for infinity (∞), so $\aleph$ puts our mind as ease as we consider the possibility of actual infinity, as opposed to facing something unnerving like 2^∞. I would argue that we only think we are really thinking about actual infinity.

Before explaining what Cantor's one-to-one correspondence or isomorph methodology means, we should start with an example to set the stage. If we want to compare the sizes of two sets of objects, one way is to count the objects in each set and compare the results. This methodology will tell us how many objects are in each set, which will allow us to compare the two sets. Another way to do this is to put the two sets of objects into a one-to-one correspondence or isomorph to assess whether the two sets are the same size, although this methodology will not tell us the size of each set. For example, if we consider the set of even numbers less than or equal to 10 and the set of odd numbers less than or equal to 10, we can arrange to two sets of numbers to show they are equal in size without knowing the size or cardinality of each set.

$$
\begin{array}{ccc}
1 & \rightarrow & 2 \\
3 & \rightarrow & 4 \\
5 & \rightarrow & 6 \\
7 & \rightarrow & 8 \\
9 & \rightarrow & 10
\end{array}
$$

For every odd number, there is a corresponding even number. When we run out of odd numbers we also run out of even numbers (we can see it), which means there is a bijection between the two sets of numbers. There is nothing "left over," which means the two sets have the same size or cardinality. In this case, the cardinality of both sets is 5, but the methodology cannot provide us this answer.

The same idea could apply to a variety of situations, such as comparing the quantity of apples in one basket with the quantity of oranges in another basket. The interesting part of this definition is that it does not require an understanding of number. If a mother has candy and plays "one for you, one for me" with her daughter, they can set the candies down in two rows to see whether they have the same quantity of candies, which they will if the number of candies is even. This methodology is useful if we think about mathematics in terms of set theory, but although numbers can be thought of as sets, they are not sets because set theory does not provide a foundation for mathematics.

As we begin our transition to Cantor's methodology in the context of infinity, the basic idea is to generalize this process such that it continues ad infinitum so that we have a systematic way to put two ad infinitum sets into a one-to-one correspondence or isomorph with nothing left over, such that there is a bijection (we cannot see it but can imagine it). For example, our two columns of odd and even numbers could continue ad infinitum. We can establish a one-to-one correspondence or isomorph to show that the set of odd numbers is equal to the set of even numbers as we build these two sets ad infinitum, but only because we know this is true on a finite level. This might sound reasonable, but this is precisely where the fun begins. Those who believe in potential infinity and those who believe in actual infinity both agree the set of odd numbers is the same size as the set of even numbers, albeit in different ways, but this agreement ends when we transition to Cantor's general methodology for all infinite sets.

According to Cantor, a set is infinite if it can be put into a one-to-one correspondence or isomorph with a proper subset of itself, such as the even numbers being a proper subset of the natural numbers. Every even number is a natural number but not every natural number is an even number, and both sets can continue ad infinitum. From the start, this is an odd way to define an infinite set because an easier and more meaningful way would be to define infinite sets as never ending, unlimited (apeiron). What do we gain from complicating the definition with proper subsets and isomorphs, especially if this definition does not apply to all infinite sets? For example, Cantor says the set of real numbers is an infinite set but claims it cannot be put into a one-to-one correspondence or isomorph with a proper subset of itself. This definition of an infinite set bakes something non-mathematical into the cake that makes Cantor's theory of actual infinity possible (more on this later). I am not suggesting that Cantor's definition is false, but we should not accept it as the only possible or complete definition of an infinite set.

What happens if we attempt to put the natural numbers into a one-to-one correspondence or isomorph with the even numbers? Clearly, the even numbers are a proper subset of the natural numbers, but does it make sense to say the set of even numbers is the same size as the set of natural numbers? If we imagine two children counting, one by ones and one by twos, we would see the following.

1 → 2
2 → 4
3 → 6
4 → 8
5 → 10
... ...

This process could continue ad infinitum but the significance is not intuitive or clear. So what? The mere fact that you can start the process of pairing two sets of numbers, which is not a mathematical operation, does not mean the two sets are the same size. The question is whether can we establish a bijection between the two sets so that no numbers are left over. This methodology works in the case of finite sets, because we can see the result, but there is no reason to assume it holds in the case of infinite sets. We will never run out of natural numbers or even numbers (hence the idea of potential infinity), but this methodology seems to confuse two things: first, the ability to pair up numbers from two sets, which we can do with any two sets of numbers; and second, the relative sizes of the two sets at any given point in the process. For example, in the above example we can pair up the natural numbers with the even numbers, but if we pair up the natural numbers and the even numbers less than or equal to 10, the set of even numbers is half the size of the natural numbers (we can see the absence of a bijection), which for Cantor is the foundational infinite set the generates all the other infinite sets. Consider the below version of the same process.

1
2 → 2
3
4 → 4
5
6 → 6

7		
8	→	8
9		
10	→	10

With this equally valid methodology of pairing up the two sets of numbers, the set of natural numbers will always be twice the size of the even numbers as the process to goes to infinity. After all, the set of even numbers is derived from the set of natural numbers, so we cannot list the first 5 even numbers unless we derive them from the first 10 natural numbers. If this process can be thought of as reaching completion, which Cantor claims is possible, then the two sets clearly are not the same size. Of interest, however, and this is where Cantor begins mixing his secret sauce, he argues that actual infinity exists and that a mysterious transformation takes place, such that the natural numbers and the even numbers are the same size when considered as infinite sets, with no explanation for what happens to the leftover odd numbers from the natural numbers.

The explanation for this is the claim that any two sets of numbers that can be paired up in a one-on-one correspondence or isomorph are the same size, even if one set grows at a faster rate than the other set. In the above example, we can pair the two sets of numbers but one set is clearly larger than the other, which means the pairing process ends before we include all the natural numbers. Of interest, when commonsense humbly suggests that the idea of pairing infinite sets violates our most basic intuitions, the Cantor camp is kind enough to advise us that our basic intuitions are flawed. Our finite minds apparently cannot grasp infinity, yet our finite minds somehow possess the ability to create the mathematics of infinity and judge whether it is right or wrong.

To reinforce this point, we can consider the same problem from a slightly different perspective:

2	→	1
4	→	2
6	→	3

8	→	4
10	→	5
		6
		7
		8
		9
		10

Again, we can see the absence of a bijection. If we think of the set of natural numbers as our home base, which Cantor does, and generate the set of even numbers from the natural numbers, then we will never be in a position to make a one-to-one correspondence or isomorph between the two sets because we will always have twice as many natural numbers as even numbers. This thought process will be important and look familiar when we discuss power sets because whereas Cantor invoked the power set to claim the power set of the natural numbers is larger than the natural numbers, he did not use the same thought process to say the set of natural numbers is larger than the set of even numbers. After all, we can pair up the natural numbers and the power set of the natural numbers and continue ad infinitum because a discrete counting process never reaches actual infinity.

For many people, this will be the decision point: if Cantor's methodology is wrong, stop the flow chart and bask in the insightful simplicity of Aristotle's potential infinity. However, we have to continue with more examples before we arrive at Cantor's most important claim, namely, that the real numbers are not countable, which means they cannot be put into a one-to-one correspondence or isomorph with the natural numbers, even though the natural numbers are a proper subset of the real numbers.

To better understand what Cantor means, rather than think about infinite sets as sets that are reached via traditional counting, they are sets that are ready-made infinite, such that adding or subtracting numbers from the infinite set does not make the infinite set more or less infinite. The key to this idea is that adding or subtracting numbers to infinite sets does not change the results when we put the infinite sets into a one-to-one correspondence or isomorph. I happen to think this idea needs clarity, but we will continue with Cantor's assumptions to see where they lead.

As I addressed in the previous chapter, we can derive the integers (the positive and negative natural numbers) from the natural numbers (the positive natural numbers) with the mathematical operation of subtraction. For every positive natural number, there is a corresponding negative natural number, which would suggest, once again, that the integers are twice the size of the natural numbers, just as the natural numbers are twice the size of the even numbers. However, if we are creative, we can put the two sets of numbers into a one-to-one correspondence or isomorph that continues ad infinitum.

1	→	1
2	→	-1
3	→	2
4	→	-2
5	→	3
...		...

Similar to the case of the natural numbers and the even numbers, this time the integers grow at twice the rate of the natural numbers. Even though commonsense tells us the integers are twice the size of the natural numbers at a finite level (because the integers have a positive and negative version of each natural number), Cantor claims a mysterious transformation takes place that makes both sets equal in size when considered as infinite sets because we can put them into a one-to-one correspondence or isomorph. The perplexing point is—to what end? Rather than use these peculiar results to reach the conclusion that actual infinity is not a viable concept, Cantor believes it strengthens the argument for actual infinity and continues down the rabbit hole.

When we transition from the integers to the rational numbers, the idea of putting the rational numbers into a one-to-one correspondence or isomorph with the natural numbers seems like an impossible task. There is a potential infinity of rational numbers between any two natural numbers (between any two rational numbers, in fact), which would suggest that the rational numbers are significantly larger than the natural numbers, so much so that the idea of establishing a one-to-one correspondence or isomorph seems like a waste of time.

To use technical language, the rational numbers are dense because we can find a rational number between any two distinct rational numbers, no matter how close they are. This means the rational numbers do not have immediate successors because they cannot be listed in order like other sets of numbers (unless we use a creative way to disentangle the density problem that I will address later). However, given that every rational number can be represented as the ratio of two integers (p/q) (we will ignore for now the set of rational numbers as decimals), we can represent the relationship visually as a matrix, with integers in the numerator (p) and the denominator (q).

p/q	1	2	3	4	5	6	....
1	1/1	1/2	1/3	1/4	1/5	1/6	....
2	2/1	2/2	2/3	2/4	2/5	2/6	....
3	3/1	3/2	3/3	3/4	3/5	3/6	....
4	4/1	4/2	4/3	4/4	4/5	4/6	....
5	5/1	5/2	5/3	5/4	5/5	5/6	....
6	6/1	6/2	6/3	6/4	6/5	6/6	....
....	....	....	....	....	....	....	....

If we expand this matrix to actual infinity to the right and down, which Cantor assumes is possible (no small assumption), we in theory should not miss any rational numbers along the way. However, the mere fact that this process continues ad infinitum does not mean it makes sense to talk about this process reaching completion. When we take the next step of putting the natural numbers and the rational numbers into a one-to-one correspondence or isomorph we use a methodology called diagonal counting (not to be confused with the diagonal argument) to show that the rational numbers are countable, meaning that they are the same size as the natural numbers.

To do this we begin at the top-left corner of the matrix and count along the diagonals of the matrix, zigzagging back and forth to count each of the rational numbers. This is illustrated below.

p/q	1	2	3	4	5	6	....
1	1/1	1/2	1/3	1/4	1/5	1/6	....
2	2/1	2/2	2/3	2/4	2/5	2/6	....
3	3/1	3/2	3/3	3/4	3/5	3/6	....
4	4/1	4/2	4/3	4/4	4/5	4/6	....
5	5/1	5/2	5/3	5/4	5/5	5/6	....
6	6/1	6/2	6/3	6/4	6/5	6/6	....
....	....	....	....	....	....	....	....

This supposedly allows us to put the natural numbers and the rational numbers into a one-to-one correspondence or isomorph, but it seems to ignore the possibility that the process continues forever (ad infinitum) or that we will run out of natural numbers before we run out of rational numbers (no apparent bijection), which is precisely what Cantor claims happens in the case of the real numbers. With diagonal counting, we begin with the rational number in the top left corner (1/1), move down one (2/1), up one and to the right (1/2), and then follow the line to zigzag diagonally to count the rational numbers.

1	→	1/1
2	→	2/1
3	→	1/2
4	→	1/3
5	→	2/2
6	→	3/1
7	→	4/1
...		...

Although the matrix and the list are somewhat intuitive, in that we can make a visual display of the rational numbers that continues ad infinitum, the notion that the natural numbers and the set of rational numbers are the same size or cardinality defies comprehension. The set of rational numbers between 0 and 1 alone is potentially infinite, which means the rational numbers have a potential infinity of numbers a potential infinity number of times (one potentially infinite set between every two natural numbers). It is true that for every natural number we can list a rational number, ad infinitum, but this applies to any two sets of numbers. However, according to Cantor, the mere fact that we can start the process of establishing a one-to-one correspondence or isomorph between the natural numbers and the rational numbers is sufficient to say they are the same size or cardinality, but he does not apply this same logic to the real numbers.

Before moving on to real numbers, we should consider these counterintuitive ideas from a different perspective. One of the important claims in this book is that the numbering system we use shapes mathematics, with the understanding that there is something more fundamental than numbers that makes mathematics possible. This is similar to the claim that the language we use shapes the way we think and communicate, with the understanding that there is something more fundamental than language that makes language possible. There are meaningful thoughts and expressions that cannot be expressed in discursive language no matter how hard we manipulate our words. In short, we should not allow the peculiarities of the numbering system to distort the way we think about mathematics.

If we look back at our matrix of the rational numbers, the first thing we should notice is that there are many repeat numbers. The numbers along the diagonal are all equal to 1, 1/2 repeats, 1/3 repeats, 1/4 repeats, and so on. Thus, if we want to use mathematics to accurately calculate the size or cardinality of the set of rational numbers and not rely on an isomorph chart that is filled with duplicate entries, we should consider the set in a more systematic way that eliminates the duplicate entries. To do this, we can once again think of the rational numbers between 0 and 1 as an iterative process—begin with all the possible rational numbers with 2 in the denominator followed by 3, 4, 5, etc.

0/2, 1/2, 2/2

0/3, 1/3, 2/3, 3/3

0/4, 1/4, 2/4, 3/4, 4/4

0/5, 1/5, 2/5, 3/5, 4/5, 5/5

0/6, 1/6, 2/6, 3/6, 4/6, 5/6, 6/6

0/7, 1/7, 2/7, 3/7, 4/7, 5/7, 6/7, 7/7

...

We can imagine this process continuing ad infinitum. The most interesting thing about this process is that every rational number along the way eventually gets duplicated—2/4 duplicates 1/2, 2/6 duplicates 1/3, 4/6 duplicates 2/3, and so on. In fact, as this process goes to infinity, some of the fractions will be duplicated an infinite number of times and no rational numbers will never get duplicated. Thus, if we want to calculate the size or cardinality of the rational numbers between 0 and 1 in a creative and intuitive way, we can think about the limit of this process:

$0/\aleph, 1/\aleph, 2/\aleph, 3/\aleph, \ldots, \aleph/\aleph$

At first glance, the Cantor camp should be intrigued with this result, which is the set of natural numbers with a denominator equal to the size of the natural numbers, which allows us to imagine a bijection. (Granted, $\aleph$ is not a number per se, but this display gives us an intuitive understanding of why the natural numbers and rational numbers between 0–1 have the same size or cardinality.) However, there are two problems. First, although this shows the limit of the process and is the only way to list all the rational numbers as fractions between 0 and 1 without duplicates, the problem is that this process "swallows up" the rational numbers into non-existence. Every rational number in this set (any natural number divided by $\aleph$) is 0. We can keep 1/2 and not delete it, and so on, but this display shows that the transition from potential infinity to actual infinity can never be made in a meaningful way.

Second, this set of numbers applies only to the rational numbers between 0 and 1, which means this process can be repeated an infinite number of times between all the other natural numbers. If we multiply the size of the rational numbers between 0 and 1 by the size of the rational numbers between 1 and 2 we should get $\aleph \times \aleph = \aleph^2$. If we allow this process for the entire set of natural numbers, in theory we should get $\aleph \times \aleph \times \aleph \times \ldots \times \aleph = \aleph^\aleph$. If this is true, it means the set of rational numbers is significantly larger than the set of natural numbers, but Cantor claimed they are the same size ($\aleph$). Thus, it appears that the non-rigorous methodology of isomorph diagrams has generated a contradiction. However, the Cantor camp would save the day with a *deus ex machina* by saying that the arithmetic operations do not apply to infinite sets! According to this new mathematics, $\aleph + 1 = \aleph$, $\aleph - 1 = \aleph$,

$\aleph + \aleph = \aleph$, and $\aleph \times \aleph = \aleph$. Cantor provides and explanation for why this is true, but he apparently missed an important point that generates a contradiction.

As I will address later, Cantor uses the power set to prove that the cardinality of the real numbers is $2^{\aleph}$, which he claims is larger than $\aleph$. Thus, even though there are many sets that can be defined as infinite, not all infinite sets are the same size. Cantor would admit that this is not intuitive, but he would turn to mathematics for proof. I recall as a Calculus student that some answers were not intuitive, but we accepted the results because the mathematics was solid. The same idea could be said to apply here, except that the mathematics is not solid. If we consider $2^{\aleph}$, which is supposedly the size or cardinality of the set of real numbers, we can ask how to think about this as an exponent, which is a special case of multiplication, which is a special case of addition. If $2 \times 2 = 2^2$, $2 \times 2 \times 2 = 2^3$, and $2 \times 2 \times 2 \times 2 = 2^4$, then $2 \times 2 \times 2 \times 2 \times \ldots \times 2 = 2^{\aleph}$ with an infinite set of 2s. (The Cantor camp might reject this claim, but as we will see, they apply the same logic to the infinite binary decimals used with the power set methodology.) However, if this same principle does not apply to $\aleph$ ($\aleph \times \aleph \times \aleph \times \ldots \times \aleph = \aleph$), we have a problem because it generates a contradiction. If we accept $\aleph < 2^{\aleph}$ and $\aleph \times \aleph \times \aleph \times \ldots \times \aleph = \aleph$, then we have the following.

$\aleph \times \aleph \times \aleph \times \aleph \times \ldots \times \aleph < 2 \times 2 \times 2 \times 2 \times \ldots \times 2$

If we consider that these two sets have the same size or cardinality, we can simplify (metaphorically speaking).

$\aleph$ ~~$\times \aleph \times \aleph \times \aleph \times \ldots \times \aleph$~~ < 2 ~~$\times 2 \times 2 \times 2 \times \ldots \times 2$~~

$\aleph < 2$

This is clearly a contradiction. In other words, either Cantor's arithmetic of infinity is not valid ($\aleph \times \aleph \neq \aleph$) or one infinite set cannot be bigger than another infinite set ($\aleph < 2^{\aleph}$). As I prove in the appendix, the same contradiction applies to infinite addition and division. If this is true, then either actual infinity is not a valid idea or all infinities are the same size; that is, infinite sets are not hierarchical.

When we make the transition from the rational numbers to the real numbers, Cantor claims the real numbers cannot be put into a one-to-one correspondence or isomorph with the natural numbers (which are a proper subset), which means there are supposedly not enough natural numbers to count the set of real numbers. Even though the set of rational numbers and the set of real numbers between 0 and 1 are potentially infinite, Cantor argues that there are more real numbers than rational numbers. He does this with what is known as the diagonal argument, which I will address in more detail later, but the important point for now is that Cantor does not explain why the one-to-one correspondence or isomorph model works for the natural numbers, integers, and the rational numbers, but not for the real numbers.

As I addressed previously on the chapter on numbers, the reason the natural numbers can be put into a one-to-one correspondence or isomorph with the integers is because the integers can be derived from the natural numbers with the mathematical operation of subtraction. Likewise, the reason the natural numbers can be put into a one-to-one correspondence or isomorph with the rational numbers is because the rational numbers can be derived from the integers with the mathematical operation of division.

The transition from the natural numbers to the integers and to the rational numbers is derived from within the numbers and the arithmetic operations, a process of organic expansion. In the case of the real numbers, however, in particular, the transcendental numbers, we cannot use the rational numbers or mathematical operations to generate the set of real numbers. No matter how much we add, subtract, multiply, or divide rational numbers, we will never generate irrational numbers. In other words, we should be skeptical about any attempts to "arithmetize" the continuum because it cannot be done within the framework of mathematics, which, admittedly, suggests that irrational numbers do not exist as complete numbers. As we will see, Cantor invokes things like least upper bounds and power sets to bridge the gap between the natural numbers and the real numbers, but these are not organic mathematical tools.

To state it another way, the real numbers are not and can never be a natural extension of the natural numbers, regardless of definitions, which is why we cannot put the natural numbers and the real numbers into a one-to-one correspondence or isomorph. It has nothing to do with the set of real numbers being bigger than the set of natural numbers or of there being more than one size of infinity. It has everything to do with the difference between the infinity of addition and the infinity of division. Granted, if actual infinity is a meaningful idea, then Cantor was right to talk about the real numbers being uncountable, but this takes us back to the original point that Cantor's theory is grounded in a philosophical idea, not mathematics. Not to mention, as I will demonstrate, if Cantor was right, it also means the rational numbers as decimals are uncountable.

Before concluding, we should consider one more example to show the problems that arise from the one-to-one correspondence or isomorph methodology. Just as we can put the even numbers and odd numbers into a one-to-one correspondence or isomorph, we can also perform mathematical operations on the individual natural numbers ad infinitum, to include exponents. For example, if we take the natural numbers and 2^n for each natural number, this process can continue ad infinitum because the numbers in the second column are generated from the numbers in the first column.

1	→	2
2	→	4
3	→	8
4	→	16
5	→	32
6	→	64
...		...

The potential infinity camp would agree that we can take 2^n of any natural number and that this process can continue ad infinitum, without making any assumptions about infinity. For the Cantor camp, however, the situation is more difficult and raises questions about infinite sets. In order for their theory to hold, they have to accept that the two sets are the same size because they can be put in to a one-to-one correspondence or isomorph. On the other hand, if we think about the set of natural numbers as a complete infinite set (and of each of the numbers in the two columns as sets), then if we take the power set of this set (2^n), the power set is larger than the set of natural numbers—that is, one infinite set is larger than another infinite set, according to Cantor. The question becomes, if the power set (2^n) of the natural numbers is larger than the natural numbers (n), then how is it possible to continue the above one-to-one correspondence or isomorph until infinity? If the power set (2^n) is larger than the natural numbers (n), then we will eventually reach the point where taking the power set will no longer be a natural number. I have no doubt that the Cantor camp will have a sober response, but the concerns are unavoidable.

Take It to the Limit

To avoid getting stuck in our NASCAR counter-clockwise driving pattern, we should step back and ask why we are wasting our time thinking about putting sets of numbers into one-to-once correspondences or isomorphs to compare their relative sizes. This does not help us solve mathematical problems or gain a better understanding of mathematics because it is not a mathematical process. As someone who has studied mathematics, when I think about large numbers or large sets of numbers, I prefer to think in terms of limits, which is a mathematical process.

If we consider the previous problem of comparing the set of natural numbers with the set of even numbers, we can do this effectively and efficiently with limits. For example, consider the following progression.

Natural numbers	Even numbers
1, 2	2
1, 2, 3, 4	2, 4
1, 2, 3, 4, 5, 6	2, 4, 6
1, 2, 3, 4, 5, 6, 7, 8	2, 4, 6, 8
...	...

As should be clear, this process can continue ad infinitum and the size of the set of natural numbers grows at twice the rate as the set of even numbers. If we want to calculate what the relative sizes of the two sets will be at some point in the future, we can use algebraic symbols to represent the problem in general terms.

$\underline{2x}$ (natural numbers)
x (even numbers)

For example, if this process continues until the set of even numbers (x) grows to a size of 135,298, then the set of natural numbers (2x) will grow to a size of 270,596. In other words, if we count to 270,596 via the natural numbers, the corresponding quantity of even numbers will be half the size. If we take the limit as x to go to infinity, then the limit of this ratio goes to 2 by using L'Hôpital's rule (the limit of a ratio is equal the limit of the ratio of the derivatives, or 2/1 = 2). This means the set of natural numbers is twice the size of the set of even numbers as x goes to infinity, just as our commonsense told us. If the two sets of numbers were the same size, the limit would be 1, not 2. That is, although the size of the set of natural numbers is always double the size of the set of even numbers, the limit process eliminates a scenario in which a mysterious transformation takes place resulting in the set of natural numbers suddenly being the same size or cardinality as the set of even numbers.

Consider the prime numbers. If Cantor is correct, the set of prime numbers and the set of natural numbers are the same size or cardinality because we can start the process of putting the two sets into a one to one correspondence or isomorph, even though the prime numbers are a small subset of the natural numbers and there is no systematic way to generate the next prime number. As we did with the previous example, we can compare the set of prime numbers and the set of natural numbers by how the set of prime numbers grows relative to the set of natural numbers. For example, we can calculate how many prime numbers exist below 10, below 100, below 1,000, below 10,000 and so on, such that we can use algebraic symbols to represent the problem in general terms. As it turns out, the Prime Number Theorem shows that the rate of growth is ln(x)/x (the natural log). As x goes to infinity, the quantity of prime grows at the rate of ln(x). Therefore, if we take the limit of this ratio as x goes to infinity (using L'Hôpital's rule), the answer is 0 ((1/x)/1), meaning that the natural numbers is larger than the prime numbers, just as commonsense told us. If the two sets were the same size, the limit would have been 1, not 0.

When we consider the real numbers, however, the situation is more complicated because we are talking about two different types of numbering systems—the infinity of addition (natural numbers) and the infinity of division (real numbers). There is no way to begin at a finite level that we can extend to infinity to determine the rate at which the real numbers grow relative to the natural numbers because the real numbers are dense and not sequential. For example, in the case of the natural numbers and the even numbers, we could begin with finite sets and take the limit as x goes to infinity to compare their relative size, just as we could with the natural numbers and the prime numbers.

In the case of the real numbers, there is no way to track how they grow relative to the natural numbers because they are infinitely larger at even the smallest level. Given this inability to define the relationship between the real numbers and natural numbers at a finite level, there is no reason to assume it makes sense to define the relationship as x goes to infinity. The more important questions relate to the relative size of the real numbers and the rational numbers, which is something I will address in more detail later. Whereas the one-to-one correspondence or isomorph methodology gives the impression that two infinite sets might be the same size, the limit process shows that this methodology is flawed because it is filled with invisible assumptions.

The Largest Natural Number

There seems to be agreement on both sides of the infinity debate that the process of counting the natural numbers discretely will never reach actual infinity, either because it is not possible or because actual infinity is not a meaningful term. Those who reject the idea of actual infinity would reject the idea of counting to infinity for obvious reasons, most notably because no matter how high we count we can always add one more number, and this additional number always makes the set larger ($n + 1 \neq n$). The process never ends, which is precisely what we mean by ad infinitum. Of interest, however, those who believe in actual infinity also believe it is impossible to count our way to actual infinity because they do not understand infinity as a big number that is achieved by counting. An infinite set can be held and manipulated or thought of as a complete whole, such as the set of real numbers between 0 and 1. For the sake of intellectual curiosity, we should consider the idea from another perspective.

Normally, when we make a list of natural numbers, we start from the bottom and work our way up to the top.

1
2
3
4
5
...

However, what if we begin from the top and work our way to the bottom? (If this sounds crazy, think about working our way from top to bottom with the real numbers between 0 and 1.) When we think of the largest natural number, we can think of it as an iterative process, just as we would think about generating an irrational number as an iterative process. For example, the largest single-digit natural number is 9, the largest two-digit natural number is 99, the largest three-digit natural number is 999, and so on. For this reason, the largest natural number must be an infinite series of 9s because any natural number that contains a number other than 9 in one of its place-value positions will be smaller than a number consisting of an infinite series of 9s. Thus, the largest natural number will look like this.

...999,999,999,999,999

We can express this number as an infinite sum, such that $\Sigma\ 9 \times 10^n$, as n goes from $0 \rightarrow\infty$. The size or cardinality of the quantity of 9s in this number is the same as the size or cardinality of the natural numbers. I have no doubt that the Cantor camp will protest this number because it harkens the end of Cantor's theory, for reasons I will explain later. Of interest, this notation appears to allow for the possibility of a natural number that is one less than the largest natural number (see below), just as there is a natural number that is one more than the smallest natural number.

However, if we are going to take the idea of actual infinity seriously, then we should have a serious discussion about what the largest natural number looks like, without assigning it a symbol or Hebrew letter to make it appear more mystical. If we can talk about an infinite sequence of numbers following the decimal, then we should be able to talk about an infinite sequence of 9s in the largest natural number. Actual infinity either exists or it does not. Take your poison. If the Cantor camp insists that having an infinite sequence of numbers after the decimal point is not the same as having an infinite sequence of numbers before the decimal, then they have admitted that there is a fundamental difference between the infinity of addition and the infinity of division, which in turn means that it makes no sense to say there are more real numbers than rational numbers, which in turns means it makes no sense to "arithmetize" the continuum.

To be clear, I am not suggesting that we can ever reach this peculiar number by counting 1, 2, 3..., but if it makes sense to talk about an infinite set as complete, such as the set of natural numbers, then we should be able to describe what both ends of this infinite set look like. Of interest, if we can postulate the largest natural number, then it also makes sense to count down from top to bottom. We will never arrive at ...3, 2, 1 via this process of discrete counting, but we can now display (not a list!) the natural numbers this way.

1
2
3
4
5
...
...999,999,999,999,995
...999,999,999,999,996
...999,999,999,999,997
...999,999,999,999,998
...999,999,999,999,999

I realize this idea might seem ridiculous or hard to grasp, but this is because the implicit assumptions of the Cantor model of actual infinity have been allowed to dominate the debate for so many years. We have been making left turns for so long because we have been driving around the racetrack counter-clockwise. In this new world, NASCAR drivers can make right turns and go clockwise. In the debate about infinity, we cannot have our cake and eat it too. If actual infinity is real, then it makes sense to talk about the largest natural number, without resorting to symbols or Hebrew letters; but if it does not make sense to talk about the largest natural number, then it raises concerns about infinity.

Just Say Nein

Sometimes a good way to understand a complex situation is to study the anomalies because these often give us clues and save us the trouble of having to study the entire spectrum of behavior to disentangle myriad variables. For example, when studying language, we can study how the use of irregular verbs develops with age to gain insights about language. A young child who starts out using the past tense verb "brought" might switch to "bringed" after he absorbs the rule of adding "ed" to make a verb past tense. Or, we can study the effects of brain lesions or traumatic brain injury to understand the brain. For example, if a person with a brain lesion or a traumatic brain injury loses the ability to empathize with people, we can conclude that the damaged part of the brain probably plays a role in our ability to empathize.

We can apply the same idea to mathematics. If we could discover any anomalies or holes in the system of base-10, place-value numbers, it might help us better understand mathematics and give us insights about the infinity debate. My favorite example is the number 9 and its cameo role in the peculiar number .999... (a series of 9s after the decimal point that extend ad infinitum). As an infinity apostate, I would say the 9s continue as long as we need them to continue, but the Cantor camp would argue that this is a legitimate number with an infinite set of 9s after the decimal point, which is possible only if the Axiom of Infinity is valid.

Anyone who was worked on multiplication charts knows about the special properties of the number 9.

1 x 9 =	09	0 + 9 = 9
2 x 9 =	18	1 + 8 = 9
3 x 9 =	27	2 + 7 = 9
4 x 9 =	36	3 + 6 = 9
5 x 9 =	45	4 + 5 = 9
6 x 9 =	54	5 + 4 = 9
7 x 9 =	63	6 + 3 = 9
8 x 9 =	72	7 + 2 = 9
9 x 9 =	81	8 + 1 = 9
9 x 10 =	90	9 + 0 = 9

The first column of the answers ascends from 0 to 9, the second column descends from 9 to 0, and the sum of the two numbers of the product is always 9. We can generalize this last rule such that the sum of the numbers of any product of 9 adds up to 9. For example, 9 x 21,486 = 193,374. If we add these numbers, we get 1 + 9 + 3 + 3 + 7 + 4 = 27. If we add these numbers, we get 2 + 7 = 9. This is enough to make mystical mathematicians or numerologists tingle with excitement, but as we will see, many of the interesting properties are a function of the fact that we use a base-10, place-value numbers. If we used a different numbering system, such as computers using a base-2 numbering system with 1s and 0s, some of these tricks would cease to exist.

A more relevant example for our analysis of infinity relates to the interesting properties of fractions of 9. Consider the following example.

1/9 = .111...

2/9 = .222...

3/9 = .333...

4/9 = .444...

5/9 = .555...

6/9 = .666...

7/9 = .777...

8/9 = .888...

9/9 = ?

If we follow the pattern, the part of us that is susceptible to magic wants to say the answer is .999..., but an arithmetic student in elementary school knows the answer is 1. One clever solution to keep the magic alive is to say both answers are correct by showing that they are the same number, as follows.

1. .999...

2. 9 x .111... (divide by 9)

3. 9 x 1/9 (express the decimal as a fraction)

4. 1 (multiply)

There are other ways to demonstrate this "proof," but before we address why this sleight of hand is invalid (we should never accept "..." at face value), we should shift gears to look at the problem from another perspective. To show how this quirk is a function of the base-10, place-value numbers, let us now shift to a base-9, place-value numbers to see the results.

1/9 = .1

2/9 = .2

3/9 = .3

4/9 = .4

5/9 = .5

6/9 = .6

7/9 = .7

8/9 = .8

9/9 = 1

This is proof that the properties of numbers are often a function of the numbering system we use, just as how we express ourselves in language is a function of the language we speak. Poetry is lost in translation. Granted, base-9, place-value numbers will introduce their own peculiarities, but we can now recognize that the .999... phenomenon is a function of the numbering system we use and the Axiom of Infinity.

What this means for mathematics might surprise some people but the = sign does not always mean what we think it means, depending on whether we accept the Axiom of Infinity. In the case of the fractions of 9 with base-10, place-value numbers, when we say 1/9 = .111..., what we are really saying is that if we divide 1 by 9, the result will be a never-ending series of 1s. From this, we cannot make the leap of claiming that this series of 1s is infinitely long and complete (the Axiom of Infinity). This leap is what I call the Ontological Argument for the Existence of Infinity (philosophy humor). In a problem with a repeating decimal, the = sign does not necessarily go both ways. We can go from 1/9 to .111..., but we cannot make the return trip of .111... to 1/9 if actual infinity is an invalid idea. A more appropriate symbol would be the implication sign of logic, such that 1/9 → .111..., which does not go both ways, or we run the risk of affirming the consequent. If we want the precision of the = sign, we should opt for 1/9.

We should take a closer look at .999... to gain additional insights. For example, what happens if we try to categorize this number? Clearly, it is not a natural number or an integer because it has numbers after the decimal point. Likewise, .999... is not an irrational number (algebraic or transcendental) because it repeats. Therefore, we must conclude that .999... is a rational number. Rational numbers by definition terminate or repeat after the decimal because the rational numbers is the set of numbers that can be expressed as p/q such that p and q are integers. If this is true, which integers p and q can we use such that p/q = .999...? The answer is none. A ratio of integers is incapable of generating .999..., so what is it? I would propose that .999... is not a number but a gap or lacuna in the base-10 place-value numbers, which is an interesting thought and a polite reminder that the numbering system we use shapes how we do mathematics, to include our hidden assumptions about infinity.

The fun does not end there, however. The idea of .999... is also important for the debate about infinity because it raises the possibility of a largest real number less than 1, which has devastating consequences for the Cantor camp. One of the most important tenants of the Cantor camp is that there exists an infinite quantity of real numbers between any two real numbers, which means there can be no smallest real number greater than 0 or a largest real number less than 1. However, if .999... is a real number, then there does not exist a real number between .999... and 1. The Cantor camp would counter that .999... = 1, but this merely kicks the can down the road.

Color by the Numbers

One of the most interesting arguments in favor of infinity has to do with potentially existing or uninstantiated properties, like colors. For example, Hume raised the scenario of a person who has seen many colors except one particular shade of blue and then uses his imagination to predict what it looks like. "Suppose therefore a person to have enjoyed his sight for thirty years, and to have become perfectly well acquainted with colors of all kinds, excepting one particular shade of blue, for instance, which it never has been his fortune to meet with."[22] In this scenario, the person goes to a paint store and sees a color array with a row of shades of blue from dark to light with the shade of blue he has never seen covered with tape. Hume concluded that if the man had a properly functioning imagination, he could probably imagine what the color should look like with a high degree of accuracy, which has important implications for epistemology and our analysis of infinity. If the salesman removes the tape to expose the shade of blue, the person probably would not be shocked by what he saw. If this is true, it suggests that we can think of the spectrum of colors as potentially infinite, with some colors existing in the world and some existing potentially (uninstantiated).

My guess is that most Eskimos would be shocked to see the full spread of colors in a paint store, but the idea here in the context of infinity is that there is a structured world out there that is beyond our perceptual grasp, either because our perceptions are limited, in the case of colors, or because we cannot perceive other things, like most of the RF spectrum. Even though we

[22] Hume, David, *A Treatise of Human Nature*, 10.

cannot achieve a complete grasp of reality, we can find clues and take steps to construct this structured world.

In the case of colors, the visible light portion of the RF spectrum runs from 430–770 THz, with every color along this spectrum a pure color, with the understanding that the individual colors of the spectrum (ROYGBIV) each cover a frequency range. For example, the color red is ~430–480 THz, the color orange is ~480–510 THz, the color yellow is ~510-540 THz, and so on, which means the colors blend into one another.

As we know from paint stores and computers, most colors are a combination of color frequencies, such as combinations of red-green-blue (RGB) or cyan-magenta-yellow (CMY). The most popular computer standard is using degrees of 0–255 of each color to generate the set of colors. For example, the three-digit code for blue for RGB is (0, 0, 255) because we use 0 red, 0 green, and 255 blue. The idea is that we can tweak the individual levels of red, green, and blue on a scale from 0–255, such that the three digit code for British Racing Green is (0, 66, 37) and the three digit code for Netflix Red is (185, 9, 11). Under this model, the number of possible colors is 256 x 256 x 256 = 16,777,216. If we were to look at each color for one second, 24 hours a day and 7 days a week, it would take about 194 days to see them all!

Before continuing, we should consider two important variables: first, the ability of the human eye to distinguish colors; and second, the number of possible colors, instantiated or not. The human eye is limited in terms of how many colors it can distinguish. Computers use the scale of 0–255 for reasons that relate to how many colors the human eye can distinguish and the properties of bytes and binary mathematics, but the important point is that the number of colors generated by this system (16,777,216) exceeds the ability of the human eye to distinguish, which means that any additional refinement would be of no value to humans. Just as technology for megapixels in cameras is limited by the resolution of the human eye, our ability to distinguish colors depends on the limits of the human eye.

However, just because a human eye cannot distinguish a color does not mean it does not exist. We could always build sophisticated equipment that can distinguish colors beyond our own capabilities, which brings us to an important question. Does it make sense to talk about the set of all colors as an infinite set? For example, rather than have a scale from 0–255, could we have a scale from 0 to infinity—continuous, not discrete? If we can have an infinite set of numbers between 0 and 1, can the colors be sliced in a similar way to achieve an infinite set of colors? The short and obvious answer is no because the RF spectrum is not infinitely divisible. Colors are frequencies or combinations of frequencies, and a frequency necessarily implies distance, which necessarily sets limits on how far something can be divided, which implies a finite number of colors. The actual number of colors might far exceed our ability to distinguish them, especially as we get down to individual wavelengths, but in the context of infinity, the size of the set of all possible colors is a tiny number.

When we talk about dividing up reality, we will always reach a natural limit, just as even the most refined and abstract vocabularies of experts or professions always reach a natural limit. Even the apocryphal Eskimos reached a natural limit for snow vocabulary. With mathematics, however, dividing a number has no impact on the number. If we divide a number 1,000,000 times it does not grow weaker or lose structural integrity, and yet this infinite divisibility of numbers gives us unreasonable ideas about how the world works. Granted, when many people talk about infinity, they are talking about something "out there" that transcends the finite world we live in, but this takes us to the realm of metaphysics, and there is no reason to assume that infinity is the only option, as I will address later. In other words, something other than an infinite set of numbers can provide a foundation for mathematics.

Conclusion

We began this chapter with a review of Cantor's non-mathematical methodology for comparing the sizes of two sets (one-to-one correspondence or isomorph). For example, if we have two baskets of fruit, one with apples and one with oranges, then if we can put the two sets of fruit into a one-to-one correspondence or isomorph, then we know the two sets of fruit are the same size or cardinality, even if we do not know the actual size. The important point is that it is a finite process that terminates, which means we can subsequently count the apples and oranges to confirm that they are the same size. Thus, at first glance, it is not clear what is to be gained by using this methodology.

As we transition to sets that are not finite, however, the idea is that we can use this same methodology to pair up one ad infinitum set with another ad infinitum set to prove that they are the same size, even if the two sets are not the same size at every finite step along the way. As is often the case, assumptions are built into this process that shape the outcome. For example, this methodology assumes that an infinite set can be thought of as complete (the Axiom of Infinity), and that a one-to-one correspondence or isomorph exists. However, we showed that Cantor's methodology has problems because the mere fact that we can start the process of pairing up two infinite sets does not mean they are the same size. For example, we can start the process of pairing up the natural numbers and the real numbers, but Cantor would argue that the two sets are not the same size. In the case of the natural numbers and the even numbers, although there is no debate that every finite set of natural numbers is twice the size of every corresponding finite set of even numbers, we are asked to believe that a mysterious transformation takes place as x goes to infinity, even though the limit process shows this is not true. The argument could also be made that there is a bijection between the natural numbers and the even numbers, such that each even number in the one-to-one correspondence or isomorph display is double the corresponding natural number, but many infinite subsets of the natural numbers do not have bijections, such as the set of prime numbers.

The literature is filled with language suggesting that the human mind is too limited to grasp infinity, but if this is true, then the human mind is also incapable of constructing a valid theory of infinity or identifying which theory is valid. It might be true that some parts of Cantor's theory of infinity are correct within his own framework, such as the real numbers being uncountable if the Axiom of Infinity is valid, but I hope that my analysis has proven that this assumption results in a contradiction.

Chapter 6: Dynamic Infinity

One of the most powerful arguments against actual infinity is the fact that we can use computers to solve any problem in mathematics using a binary numbering system (1s and 0s) without infinity. On the other hand, set theory with infinity has failed to solve problems we know are true, which means set theory is complete. Computers present our work to us in base-10, place-value numbers, but the computer itself works in 1s and 0s, similar to how Cantor works with infinite binary decimals. The problem for the Cantor camp is that computers cannot process infinity while solving problems and yet the problems get solved, which raises an important question: What purpose does infinity serve? If infinity was required for mathematics and computers cannot use infinity, then there would be some problems that computers could not solve, but this is not the case.

The Cantor camp insists that infinity should be thought of as a mathematical problem with rigorous definitions, not a metaphysical problem (even though they make metaphysical claims about imposing points onto a continuum), but it appears to serve no purpose on practical grounds. As a Calculus student, I believed in infinity because I was a member of the Cantor camp, but the truth was the existence or non-existence of infinity had no bearing on how I solved problems or which problems I could solve, even though the problems were phrased in the language of a dynamic, infinite process. However, people who converted to the Cantor camp could not suddenly solve problems they could not solve before, and people who lost the faith did not suddenly lose the ability to solve problems they could solve before.

Therefore, rather than continue to focus on theory, I thought it was a good time to do some actual mathematics to see how the theory stacks up in the real world. For me, nothing is more real world in mathematics than Calculus. Mathematics students spend most of their formative years studying the tools that allow us to solve Calculus problems (arithmetic, algebra, geometry, trigonometry) but most students never make it to Calculus and never enjoy the insights it provides, which probably explains why so many people are turned off by mathematics. Imagine if we attended band class and practiced only scales and chords, but never got the opportunity to play actual songs in an orchestra. Most people would quit. Granted, we can use arithmetic and algebra to solve real world problems, but nothing on the scale of Calculus, which allows us to land a man on the moon and other great accomplishments. Calculus is where it all comes together. Applied mathematics can go in other directions, such as probability & statistics or engineering, but pure mathematics tends to go in the direction of theory.

The most important concept in Calculus is the limit, often symbolized by $x \to \infty$, which means, "as x goes to infinity." Calculus students are asked to repeat this mantra over and over until it is accepted as an article of faith, as if this is really what is accomplished. For example, if we consider the fraction $1/x$, then as x goes to infinity, the value of the fraction goes to 0. However, we could also say, "as x goes to an obscenely large number." As long as x is a natural number, however big, the fraction will never equal 0, but the concept of a limit allows us to solve problems that supposedly could not be solved otherwise. The most important thing to remember is that this is an intellectual leap, not a quantitative leap. We never actually plug in large numbers to see what happens. When we study Calculus, we learn to think differently, in ways the uninitiated cannot understand. Even though this different way of thinking helps us solve problems, it does not change the fact that when we solve problems in Calculus, we never use infinity.

Points of Continuity

When I was studying Calculus and solving complex problems, I was always perplexed by the caveat for some of the problems that such and such a function was "continuous." As opposed to what? I was not concerned with theory at this point, but the way I thought about it was that I could plot a continuous function on a graph with a continuous line without lifting my pen, like a needle running uninterrupted along the groove of a record to play a steady stream of music, analog not digital—continuous. This is not the best way to think about it, but it gets us started down the right path. The truth is, the line on the graph was merely a visual display of the function, to add to our intuitive understanding of the problem, so it was not clear to me why we had to make any assumptions about it, and there was no reason rethink mathematics because of it.

The few times I inquired about the need for the continuity caveat I was told there were "no gaps" on the graph. When asked what this meant, because I did not see any gaps on the graph when I used my pen to draw the function, I was told that we could plug any number into the function (x) to generate a result (y), from for the entire range of possible x values—no gaps. Likewise, if we know the range of y values for a particular function, we know that every number within that range of y values would have at least one x value that would generate the y value. As I dug deeper I learned that this caveat was another way of saying the range of values along the x and y axes was the real numbers, meaning that we could include irrational numbers like $\sqrt{2}$ and π—no gaps. We could never actually plug these numbers into a function, because we would continue writing non-repeating decimals for eternity (and therefore it was not an option), but we were allowed to use the symbolic forms and ignore the reality of what we were really doing.

With most of these problems, we were solving for the derivative or the integral, often to resolve practical problems about acceleration or cooling. I could not imagine a scenario in which we would want to know how fast a car is moving after accelerating for π seconds or the change in water temperature after $\sqrt{2}$ minutes, but I saw no reason to object and could only assume that others had found situations to use these numbers. However, as my interests shifted from applied to theoretical mathematics, to include the philosophy of mathematics, I realized the issue had less to do with continuity and more to do with points and ranges, in terms of clarifying the core issue. Not only that, the issues had to do with a peculiar type of point—the dimensionless points of the real numbers. After all, if we were to solve a function for every rational number with a point with extension, we would eventually fill the line with points such that it appeared to be continuous—no gaps.

However, and this is the interesting part, the Cantor camp would claim that even if we made a dimensionless point at every rational number on the continuum (an infinite set of points, if such a thing exists), there would be many gaps that had to be filled with real numbers. Of course, an infinite set of dimensionless points would not fill any percentage of a continuum because 0 + 0 + 0 + ... + 0 = 0. To demonstrate this fact, consider the following continuum that contains a black and a gray range separated by a point:

For this example, we will assume that the line has a length of 1, which is always an arbitrary decision. Next, we will assume that the black range and gray range each have a length of 1/2, which generates a point between them. (Please note that the point did not exist prior to imposing the black and gray ranges onto the line.) Therefore, if we add the black and gray ranges (1/2 + 1/2 = 1), this means the point that separates them is dimensionless because the line has a length of 1. We can continue this process:

In this case, we have two black and two gray ranges that add up to 1 (1/4 + 1/4 + 1/4 + 1/4 = 1), which means the three points separating them are dimensionless. We do not need to continue this line of thought to state the obvious: an infinite set of dimensionless points never fills the continuum, even the set of real numbers. A finite continuum consists of finite ranges with magnitude, not dimensionless points. In fact, it does not make sense to talk about dimensionless points until we define the ranges, and the only way to establish a second point distinct from the original point is to move a measureable distance. If we take two dimensionless points along a finite continuum and press them together like books, we really have one dimensionless point, which means a finite continuum cannot consist of an infinite set of points.

Continuing with the idea of dimensionless points, one day I noticed that my cell phone felt thicker than usual. Confused, I checked it to see that I had 25 apps running in the background. I could scroll from one app to the next to select which one I wanted to use, but I opted for the "close all" option, closed all the apps, and returned my cell phone to its normal width. All kidding aside, the idea of a dimensionless point or slice make sense, such as adding multiple dimensionless layers to a digital photograph or the black and gray lines. We could have thousands or even millions of layers on a digital photograph and we would not need a special monitor to accommodate them because the additional code to add the layers merely extends the code and does not have dimension. Thus, the idea of a dimensionless point seemed reasonable—why not? If I can do it on my computer I can do it on my graph paper in Calculus. However, if points have dimension, however small, then a finite continuum will accommodate a finite quantity of points, and the finite points will fill the continuum.

I previously addressed the difference between the infinity of addition (quantity), which is the mathematics of rational numbers, and the infinity of division (the continuum), which is the mathematics of real numbers. One of the most important differences was that although the mathematical units of quantity are grounded in reality and independent of the human mind (one rock, one cat, etc.), the mathematical units of the continuum are arbitrary and created by the human mind. If someone were to point to a continuum on a computer (perhaps a mathematics program for drawing functions) and ask us how long it is or how many points we see, we could only shrug and say we have no idea. However, if someone were to ask us how many inches the continuum is, we could measure it to answer the first question but still not answer the question about points. We would recognize that there is nothing inherently quantitative about the continuum—it has no content.

The Cantor camp would have us believe that every finite continuum has the same quantity of dimensionless points (the set of real numbers), regardless of the length or regardless of what unit of measurement we use. The peculiar thing is that depending on what arbitrary unit of measurement we impose on the continuum, the same dimensionless point that represents 1 in the first unit of measurement could represent $\sqrt{2}$ in the next unit of measurement. We could do this by relabeling the old $\sqrt{2}$ point as 1. What this suggests is that the mathematics of the continuum is less about real world quantity and more about an abstract sense of the relative position of dimensionless points. Some schools of thought think of mathematics as nothing more than a system of abstract symbols and rules that have nothing to do with the real world, and perhaps now we are beginning to see why. I will raise this again when we address the idea of counting the real numbers.

If that was not confusing enough, we should also consider the idea of a dimensionless point in more detail. As I mentioned before, we can think about a dimensionless point, such as the point between the black and gray ranges, but what are the implications for the real world, or at least for a world where we talk about things like continuums and ranges (extension)? As a mental exercise, we can imagine 10 dimensionless points or slices numbered 1 - 10 and lined up and touching on both sides like billiard balls—continuous. We can imagine a software program that allows us to remove a particular point or slice from the row and return it. So far, not a problem, but what happens if we place these 10 dimensionless points or slices on a continuum and measure the distance from the first to the last. If the width of a dimensionless point or slice is 0, then we should measure the distance from 1–10.

$$0 + 0 + 0 + 0 + 0 + 0 + 0 + 0 + 0 + 0 + 0 = 0$$

We can also add the distance covered by an infinite set of dimensionless points.

$$0 + 0 + 0 + 0 + 0 + 0 + 0 + 0 + 0 + 0 + 0 \ldots + 0 = 0$$

Once we understand that the infinity of Cantor hinges on the idea of dimensionless points, it should be clear that the center does not hold. The idea of dimensionless points with extension along a finite continuum violates the law of contradiction because points either have or do not have dimension, and neither option allows for the set of real numbers to be distributed evenly along a finite continuum. Again, we can imagine something without dimension, like a point, but the Cantor camp insists that their dimensionless points add up to a number larger than 0 along a continuum.

To show how peculiar this thinking becomes, consider the following argument:

1. If we add up the ranges of dimensionless points for the rational numbers, we travel a distance of 0.

2. Given that a continuum with range consists of an infinite set of dimensionless points, the numbers other than the rational numbers (the irrational numbers) must make up the difference along the continuum.

3. Therefore, the irrational numbers fill the gaps and are infinitely larger than the rational numbers.

The error of this logic stems from the idea that as we plot our mathematical functions onto a continuum that we can in turn rely on the continuum to tell us something about mathematics that cannot be proven or demonstrated within mathematics.

The bottom line is that the mathematics of dimensionless points and the mathematics of range and extension are incompatible. The best way to resolve this debate is to recognize that things like lines, time, and space do not consist of points and that continuums are a perceptual and rational way of thinking about the world. Lines, time, and space might indeed be continuous, but this does not mean mathematics (arithmetic) is as well. We use mathematical systems to impose things like numbers and points onto reality to solve problems and to understand the world. When we enter the world of ranges and extension, however, we necessarily presume a minimal distance that makes movement along a continuum possible. When we impose the mathematics of quantity onto the continuum, the only way to "fill the gaps" is to presume a minimal distance that will allow us to solve the problems we need to solve with the right degree of precision. Of course, if a minimal distance must by definition be presumed in order to solve problems on the continuum, the idea of infinity dissipates, like a sparkler on Independence Day that makes us smile before fizzling away.

Infinitesimals

For reasons that are not clear, mathematicians through the ages have been predisposed to begin with the assumption that a line consists of an infinite number of points. We cannot see these points, but the idea is that if we take a glowing point and move it quickly through the dark of night, like waving the glowing end of an incense stick, it will generate a glowing line. The problem with this satisfying image is that it is not true because if we cannot see a dimensionless point, we cannot see a dimensionless line. In other words, if we can see the glow of the point and use it to create a glowing line, it means the point is not dimensionless. As surprising as this might seem to have to repeat, dimensionless means dimensionless, which in mathematics or physics means lacking extension.

If we continue this line of thinking, in theory we could take a dimensionless line resting on a desk, hold it at both ends with our fingertips, and raise it quickly so that the line traces the shape of a two-dimensional square. The idea here is that if a line consists of an infinite number of points, then a square consists of an infinite number of lines. However, the same problem persists. If we cannot see dimensionless points, we cannot see dimensionless lines and no amount of them, even an infinite set, will ever constitute a square. Moving on to the third dimension, if we take a square resting on a desk (with a height of 0) and raise it quickly, it will trace the shape of a three-dimensional cube. The idea here is that if a line consists of an infinite number of points and a square consists of an infinite number of lines, then a cube consists of an infinite number of squares. We can certainly imagine this in an ideal realm, but the idea of trying to reduce cubes, squares, or lines to actual infinite sets of squares, lines, or points in the ideal or real world generates a contradiction. There is no way to evenly distribute an infinite set of points along a finite continuum, and so on.

Up to this point, we have discussed the idea of dimensionless points as if they are not real or exist in a Platonic realm. However, with mathematicians like Galileo and Wallis, the idea of infinitesimals was considered in a material or scientific way. For example, whereas things like rope or wood are held together with a finite number of fibers, in the case of more continuous substances like marble or steel, they postulated that what held them together was an infinite number of irreducible components, like atoms, which cannot be further divided—infinitesimals separated by an infinite number of empty spaces. (The problem with this distinction is that we can ask what holds the individual rope and wood fibers together.) This attempt to think about mathematics in a scientific way seems like a positive development, but, once again, it is not clear what is gained by proposing an infinite set. Why not simply accept that the smallest, irreducible components of the universe have dimension? Atoms and the components of atoms have dimension, which necessarily means that every finite object contains a finite number of atoms. In fact, we can make these calculations with precision—the size of an atom, the weight of an atom, etc.

Historically, the rise of the mathematics of Galileo and Wallis was at least partly in response to the dominance of deductive Euclidean geometry, which was studied vigorously and promoted by the Jesuits of the Catholic Church. "The fight was over the face of the modern world. Two camps confronted each other over the infinitesimal. On the one side were ranged the forces of hierarchy and order—Jesuits, Hobbesians, French royal courtiers, and High Church Anglicans. They believed in a unified and fixed order in the world, both natural and human, and were fiercely opposed to infinitesimals. On the other side were comparative 'liberalizers' such as Galileo, Wallis, and the Newtonians. They believed in a more pluralistic and flexible order, one that might accommodate a range of views and diverse centers of power, and championed infinitesimals and their use in mathematics."[23] If so, such a misunderstanding was an unfortunate footnote in the history of mathematics. The scientific revolution was all about the pursuit of truth and a unified theory of the universe.

Historically, the ancient Greeks shifted their emphasis from arithmetic to geometry in response to the problem of irrational numbers, but the pendulum has swung the other way with the modern world, back to arithmetic (this time embracing irrational numbers and infinity) and away from geometry. However, as this book has made clear, the two options are not mutually exclusive—the infinity of addition and the infinity of the continuum. The Jesuits were perfectly correct in their claim about the irrefutability of Euclidean geometry (in its own context) and thus had no reason to fear the rise of infinitesimals as a threat to the natural order. They were not justified, however, in using the full force of the Catholic Church to censor or punish proponents of infinitesimals. Likewise, Galileo, Wallis, and others were perfectly correct in their claim that there were new and creative ways to calculate the areas of shapes or solve other types of problems. Their methodology often relied on mathematical induction, not deduction, so there is no real conflict between the two methodologies. Galileo and Wallis never denied the validity of deduction or geometry. Solving geometric proofs and polynomial equations are two different types of problems. The problems arise when we attempt to "arithmetize" the continuum by blending the two realms, which I have argued in this book is not possible and leads to contradiction.

[23] Alexander, Amir, *Infinitesimal: How a Dangerous Mathematical Theory Shapes the Modern World*, 8.

To show the problem of infinitesimals, consider the following rectangle:

If we suppose that the height is 2 and the width is 4, then the area is 8. However, if we employ infinitesimals, the results generate a contradiction. Rather than multiply the height by the width to calculate the area, we can think of the rectangle as consisting of an infinite set of dimensionless lines, either vertical or horizontal. However, whether we go vertical or horizontal, both contain the same quantity of dimensionless lines (the set of real numbers), even though the two sides are not the same length. Therefore, if we make vertical lines, the area of the rectangle equals the area of each line ($2 \times 1/\infty$) multiplied by the quantity of lines (∞), which is $2 \times 1/\infty \times \infty = 2$. However, if we make horizontal lines, we use the same formula but the results are not the same ($4 \times 1/\infty \times \infty = 4$). Therefore, one of the fundamental claims of Cantor (that every finite continuum has the same cardinality of points) is shown to generate a contradiction.

The most important point of this historical conflict is the recognition that how we think about mathematics and infinity has important consequences for society and politics, rightly or wrongly. After all, this age saw of the rise of Martin Luther and the Protestant Reformation, which forced many scientists and mathematicians to take sides and continues to shape the world today. There is no doubt that how we think about infinity can have theological implications, but we should not allow the way we do mathematics to shape our politics, just we should now allow our politics to shape the way we do mathematics. As we will see in the next section, the methodologies we use in mathematics need not have metaphysical implications. Just because we frame a problem as an infinite process to help us understand or solve a problem, it does not mean that the infinite process exists. For example, even though we can think of a shape with curved sides as consisting of an infinite number of lines to help us calculate the area, this does not mean the shape consists of an infinite number of lines.

The Algebra of Calculus

When Calculus textbooks use terms like point and continuity, the implicit assumption is that infinity is real and is required to solve problems, which this section will prove is not the case. With Calculus we transition from static infinity to dynamic infinity—a scenario in which something transforms via an apparently infinite series of steps to arrive at an end state. Broadly speaking, Calculus solves two types of problems: derivatives and integrals. Beginning with derivatives, they are used to calculate the slope of a curved line at a specific point. Derivatives have applications that go beyond calculating the slopes of curved lines, such as calculating the speed of a vehicle that is accelerating, but for now we will focus on the mathematics of solving a derivative.

Normally, the slope of a straight line is calculated by taking the rise over the run. For example, the slope of the function $y = 2x - 1$ is always 2, because every time we move 1 place to the right, the line rises by 2; it goes up 2 and over 1. The slope of this line is the same at every point on the graph, which means the function is a straight line.

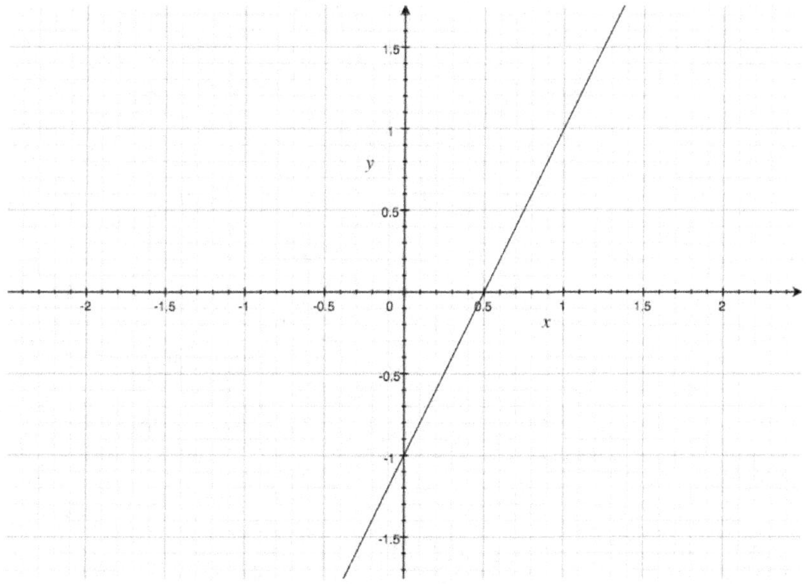

However, when we consider a function that is not a straight line, like the function $y = x^2$, we cannot use the normal rise over run formula to calculate the slope of a line at a point because the line is curved, which means the slope of the line is not static throughout the function and is always changing.

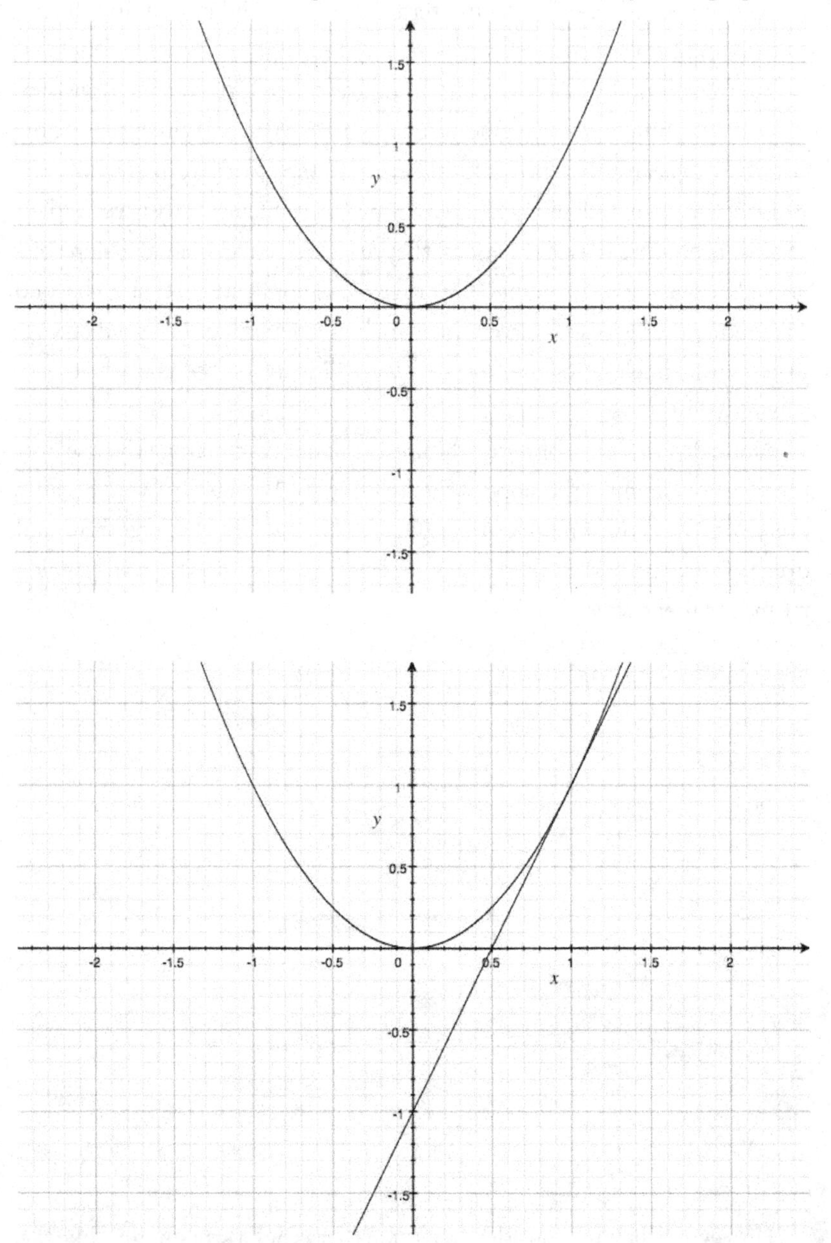

If we calculate the function when $x = 1$ (1^2), the answer is 1, which is easy enough. However, if we draw a line to estimate the slope of the line at the point ($x = 1, y = 1$), as shown on the previous page, there is no way to draw a line that is perfectly tangential to the curved line by hand with a pencil and ruler, because the line is curved. We can estimate the slope, but we will require the tools of Calculus to derive a precise answer that will meet the rigorous standards of mathematics.

One way to estimate the slope of the curved line when $x = 1$ is to use the rise over run formula by solving $y = x^2$ for $x = 1$ and $x = 2$ and then draw a straight line to connect the two points. Of importance, we will refer to the distance between $x = 1$ and $x = 2$ as the "change of x," or Δx, using mathematical symbols. In this case, as we move from $x = 1$ to $x = 2$, $\Delta x = 1$. Likewise, as we move from $y = 1$ to $y = 4$, $\Delta y = 3$. This straight line crosses the curved function at (1, 1) and (2, 4), such that the rise is 3 (4 - 1) and the run is 1 (2 - 1), which means the slope is 3 (3/1). However, it should be obvious that this estimate is not accurate. Next, we could reduce our Δx from 1 to 0.5 to calculate the slope of the straight line that passes through the curved function at (1, 1) and (1.5, 2.25). In this case, the new slope is 2.25 (1.25/0.5) because we have a rise of 1.25 (2.25 - 1) and a run of 0.5 (1.5 - 1). Next, we could reduce our Δx from 0.5 to 0.25 to calculate the slope of the straight line that passes through the curved function at (1, 1) and (1.25, 1.5625). In this case, the new slope is 2.25 because we have a rise of 0.5625 and a run of 0.25.

We can continue this iterative process to see what happens as Δx approaches 0.

Δx	Slope
1	3
0.5	2.5
0.25	2.25
0.125	2.125
0.0625	2.0625
0.03125	2.03125
0.015625	2.015625

Thus, as Δx gets closer 0 (as we get closer to drawing a line that touches $y = x^2$ at precisely one point), the slope for $y = x^2$ when $x = 1$ seems to be converging on 2, such as the graph for $y = 2x - 1$. In fact, if the pattern continues, it seems that we could simply change Δx to 0 to get a slope of 2, but we are not allowed to divide by 0. Thus, it appears that we can never be 100% sure about the slope of the line when $\Delta x = 0$ because we have an ad infinitum process of getting the Δx as close to 0 as possible without dividing by 0, which is how the mathematics of limits is often understood. Thus, textbooks teach that calculating the derivative of a function (which allows us to calculate the slope of a curved line at a specific point) is the result of a limit process.

But is this really the case? Sometimes contrary evidence is staring us in the face and we do not see it. If we take the graph of the function $y = x^2$ and use a pencil and a ruler, we can estimate the slope of the function $y = x^2$ when $x = 1$. We will not necessarily get the perfect answer, probably something close to 2, but if we believe that what appears on the graph reveals what is happening in the guts of the function (the graph is a visual display of the function, not the other way), our intuition should tell us something is wrong.

It is true that we cannot divide by 0, because weird things happen when we do, such as gaps or asymptotes. However, if this was the case, we should see strange things happen to the estimated slope as we approach $\Delta x = 0$. However, this is not the case. On the contrary, the process seems to be converging to a slope of 2 in a steady and gradual way, almost as if the possibility of $\Delta x = 0$ does not matter. Therefore, rather than run more iterations, we should treat the Δx process itself as a mathematical function to see whether we can solve it with algebra in such a way that the ideas of limits or infinity are no longer required. In this case, using the Δx notation, the formula for the slope of a line at any given point is rise over run, which can be expressed as:

Rise: $\underline{f(x + \Delta x) - f(x)}$

Run: Δx

Thus, from the beginning we see that plugging in 0 for Δx will not work. If we plug in the formula $y = x^2$ and do the math, we get:

1. $\dfrac{(x + \Delta x)^2 - x^2}{\Delta x}$
2. $\dfrac{\cancel{x^2} + 2x\Delta x + \Delta x^2 - \cancel{x^2}}{\Delta x}$
3. $\dfrac{\Delta x^2 + 2x\Delta x}{\Delta x}$
4. $\dfrac{\cancel{\Delta x}(\Delta x + 2x)}{\cancel{\Delta x}}$
5. $2x + \Delta x$

When $\Delta x = 0$, the slope of the line at any point on the function $y = x^2$ is $2x$, like our original line $y = 2x - 1$, which is a general solution for all the possible values of x, without the need of an iterative process or estimates with a ruler and pencil. The reason we can allow $\Delta x = 0$ is because Δx is no longer in the denominator. Thus, as our intuition correctly sensed after several iterations, the slope of the line of $y = x^2$ when $x = 1$ is 2, exactly 2. What began as an apparent ad infinitum, iterative process was reduced to simple algebra.

The most important fact is that the Δx canceled out because there is a Δx in the numerator and in the denominator, and they cancel each other out each step of the way. This means we can calculate the slope of any curve at any point by using nothing more than high school algebra, without imaging an infinite process. Granted, we framed the problem in terms of an infinite process (as Δx approaches 0), but limits and infinity played no role in our solution. The idea of limits and infinity need never enter our minds, which was why anyone can solve Calculus problems with a limited understanding of infinity.

The good news is that we can generalize these results beyond $y = x^2$. If we focus on polynomial equations, the derivative of any formula is as follows: the derivative of $ax^n = nax^{n-1}$, such that the derivative of $3x^3$ is $9x^2$. Admittedly, Calculus is more complex and many functions are not so nice and tidy but the fact that so much of this can be generalized with algebra is important as we shift gears from derivatives to integrals.

Whereas we use differential calculus to calculate the slope of a curve, we use integral calculus to calculate the area under the curve, which has applications that go beyond calculating the area under a curve. Continuing with $y = x^2$, what if rather than calculate the slope of a tangent at a point along the curve we decided to calculate the area under the curve? We know the formulas for calculating the areas of triangles, quadrilaterals, and other shapes with straight sides, but what happens when one side is curved? For example, if we decide to calculate the area under the curve $y = x^2$ from 0 to 2, we could estimate the area by drawing a right triangle, with the hypotenuse connecting the points where $x = 0$, $x = 2$, and $y = 4$. The area of this triangle is 1/2 x 2 x 4 = 4. However, given that the hypotenuse of the triangle is always above the curve, this estimate is not accurate. Not to mention, there does not seem to be an iterative process whereby we can get closer and closer to true area.

For this task, rectangles seem like a better option. To avoid having the entire curve above or below the line, which makes for inaccurate results, we can draw rectangles such that the gaps of coverage are offsetting. For example, given that we are calculating the area under the curve between x = 0 and x = 2, we can construct two rectangles, one with a width of x = 0 to x = 1 and another with a width of x = 1 to x = 2. Regarding the height, rather than calculate the height based on x = 1 or x = 2, which will result in rectangles that are too big, we can select x = 0.5 and x = 1.5, which will create offsetting areas for better accuracy.

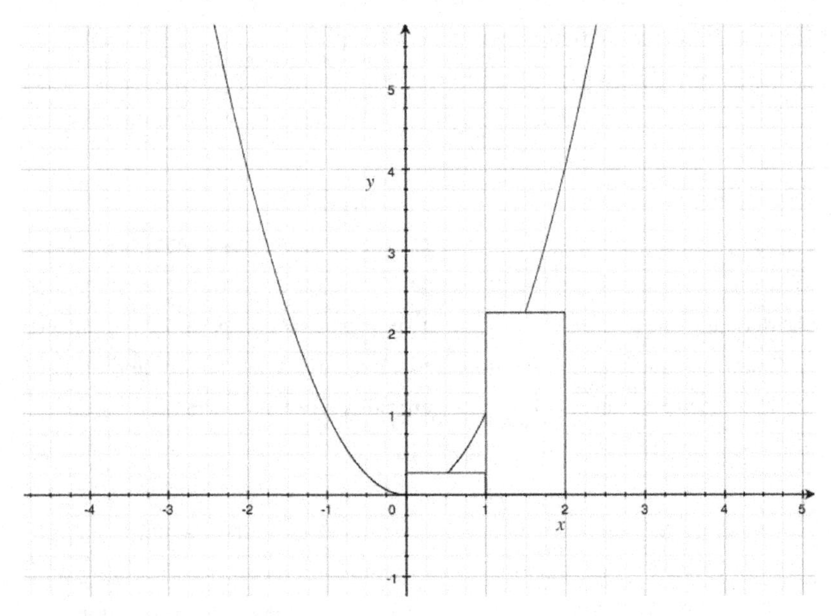

The area of the first rectangle is 1 x 0.25 = 0.25. The area of the second rectangle is 1 x 2.25 = 2.25, for a total area of 2.5, which seems more accurate than our original triangle estimate of 4. The benefit of using rectangles is that we can use an iterative process of making narrower rectangles. We can next create four rectangles, each with a width of 0.5, again using the mid-point to calculate the height of the rectangle. The areas of the four rectangles are as follows: (0.5 x 0.0625 = .03125) + (0.5 x 0.5625 = 0.28125) + (0.5 x 1.5625 = 0.78125) + (0.5 x 3.0625 = 1.53125), for a total of 2.625.

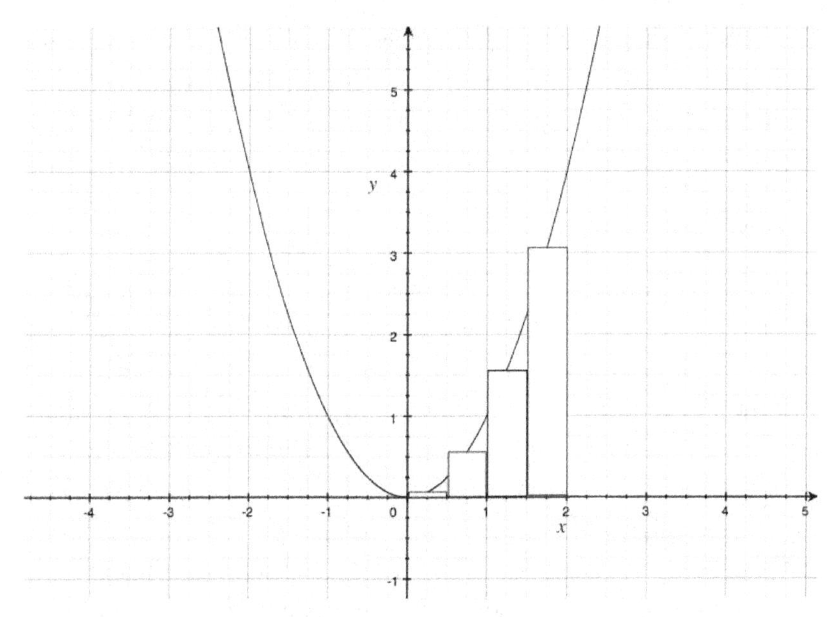

With these two calculations, 2.5 and 2.625, it is difficult to estimate where this iterative process is going, but we can create 8 rectangles, 16 rectangles, or 32 rectangles to get a more accurate estimate for the area under the curve, until the rectangles are so narrow that the human eye cannot discern the width, as x goes to infinity.

Just as we can skip this iterative process with derivatives to solve the problem with algebra, it turns out we can do the same with integrals, which brings us to one of the most important findings in the history of mathematics. According to the Fundamental Theorem of Calculus, derivative and integrals are inverse functions. That is, if we take the derivative of a function, such as $y = x^2$, which is 2x, then the integral of 2x is simply the original function x^2. Thus, given that we can solve derivatives with algebra and given that derivatives and integrals are inverse functions, we can calculate integrals with algebra, with no need for limits or infinity.

In the case of $y = x^2$, the integral is $y = 1/3x^3 + C$ (a constant) because if we take the derivative of $y = 1/3x^3 + C$ we get $y = x^2$ (the derivative of a constant is 0). Thus, to calculate the area under the curve $y = x^2$ from 0 to 2, we plug the values 0 and 2 into the formula for the integral:

$1/3(2^3) - 1/3(0^3) = 8/3 - 0 = 2\ 2/3$

With this simple algebraic formula, we can calculate the exact area under a curve without resorting to an ad infinitum iterative process, which is additional proof that we need not make any assumptions about infinity to solve Calculus problems.

The Power Set

Set theory is concerned with the mathematics of infinity, in particular, infinite sets, and does this by assuming the existence of an infinite set of the natural numbers with the Axiom of Infinity. In fact, the Cantor camp claims that the mathematics of the real numbers requires the existence of infinite sets and is not possible without them, which might be true. Given set theory's place of prominence in mathematics and the influence it has wielded over the last 100+ years, I understood that my book would be incomplete without addressing what it is, what is says about infinity, why I disagree with it, and how it can be shown that Cantor's argument fails within his own assumptions. If we can show that Cantor's proofs for the cardinality of the rational numbers and the real numbers are wrong and that his assumptions and definitions generate contradictions, it should go a long way to ending the debate about infinity.

Cantor developed set theory in the late nineteenth century, when he devised his famous diagonal argument, but his system was flawed (it was called "naïve") and was replaced with axiomatic set theory, known as Zermelo-Fraenkel Choice (ZFC). To date there is still no agreement on exactly how set theory should be defined or understood—it is still a work in progress and is still unable to solve some mathematical problems we know are true. The reason Cantor's theory was deemed "naïve" was because his system could not handle the idea of a "set of all sets" without generating a contradiction. To address this conundrum, set theory introduced the idea of a class, which is beyond the scope of this book. The proponents of set theory are still arguing about adding new axioms to "fill the gaps" of set theory while mathematicians continue to solve real problems.

In geometry, we use self-evident axioms to prove the truths of geometry because—this is important—it is they only way to solve problems in geometry. The axioms of geometry actually provide the foundation for geometry. In the case of set theory, however, which deals with sets, not geometric shapes, there is no need to invoke axioms because we can solve arithmetic problems, to include algebra, trigonometry, and calculus, without set theory or the axioms. On top of that, at least one of the axioms of set theory is not self-evident, notably, the Axiom of Infinity. For over 2,000 years, virtually all of the greatest minds in history dismissed the idea of actual infinity, as did many of Cantor's colleagues, so skepticism about Cantor should not be viewed with skepticism. Despite the apparent victory of Cantor and the claim that the issue is settled, my research left me with sufficient evidence that the issue is not settled and that many reasonable people are still willing to dismiss the idea of actual infinity.

Moving on to the Axiom of Power Set (Cantor's secret sauce), it states, innocently enough, that for any given set, the power set exists, and secondly, that the power set is always larger than the original set. As innocent as it sounds, it is one of the most important foundations for Cantor's theory of infinity, and therefore merits our attention. In fact, Cantor's theory sinks or swims based on the power set. At first glance, there is no reason to invoke such an axiom, unless someone plans to invoke it in a place that might surprise us or defy intuition, such as infinity.

If we think about mathematics in terms of sets, we can think about the subsets of sets or the set of all subsets of a set. For example, consider the following set: (2, 4, 7). The size or cardinality of this set is 3 because it contains three elements, even though the number 3 is not a member of the set. Along these lines, the two sets (3, 5, 9) and (2, 4, 7) are the same size or cardinality (3) because they can be put into a one-to-one correspondence or isomorph with each other (we can see the bijection).

3 → 2
5 → 4
9 → 7

If we consider the set of all subsets or the power set of (2, 4, 7), which means all the possible combinations of the three numbers, we have the following set: [(0), (2), (4), (7), (2, 4), (2, 7), (4, 7), (2, 4, 7)]. In this case, the power set of a set with 3 elements has 8 sets (not 8 numbers), and the power set includes the null set (0) and the original set (2, 4, 7). If we repeat this process enough times we can generalize our findings to prove that for any given set with n elements, the power set has 2^n sets (2 x 2 x 2 x ... x 2), meaning that if the cardinality of a set is n, then the cardinality of the power set is 2^n. (We should note that 2^n is an arithmetic operation and constructing a power set is not.) The most important point about power sets is that the original set and the power set do not have the same cardinality and therefore cannot be put into a one-to-one correspondence or isomorph (no bijection). For any n, the power set is larger than the original set.

As I have showed in this book, what looks simple and reasonable with finite numbers or sets can mysteriously transforms into a conundrum when we enter the realm of infinity. To cut to the chase, if infinite sets are real (per the Axiom of Infinity) and infinite sets have power sets (per the Axiom of Power Set), then if the cardinality of the set of natural numbers is $\aleph$, then the cardinality of the power set is $2^{\aleph}$. (According to Cantor, the arithmetic operations do not function on infinite sets, but this appears to be the exception.) As it turns out, and this is the key, Cantor argued that $2^{\aleph}$ is the cardinality of the real numbers ($\aleph_1$), which is why he claims the real numbers cannot be counted and are larger than the set of natural numbers. Most of the mystery of Cantor's theory boils down to this unusual claim. If I have been successful in this book, my hope is that the reader will at least pause. Just to be complete, however, we should explain why this argument fails.

First, this methodology shows the challenges of thinking in terms of actual infinity in a consistent manner. Consider the following display showing the cardinality of a set and the cardinality of the power set, which I raised previously.

Set	Power Set (2^n)
0	1
1	2
2	4
3	8
4	16
5	32
6	64
7	128
8	256
...	...

Every step of the way, the cardinality of the power set is larger than the cardinality of the original set, just as it would be for 3^n or 10^n. In fact, the power set grows exponentially by a power of 2. However, this process should continue ad infinitum because even though the cardinality of the power set is larger than the original set, it is nonetheless countable (the results of an arithmetic operation), just as the set of squares and cubes is countable. This means we can always add a finite number to the original set to equal the cardinality of the power set, which means that even though the original set and the power set are separated by the exponent 2^n, we can always bridge the gap with addition or multiplication. This is important because, according to Cantor, any number that is added to the set of natural numbers (ℵ) does not increase the cardinality of the natural numbers—that is, $\aleph + 1 = \aleph$. However, Cantor claims that if we add enough numbers or multiply enough numbers to the set of natural numbers, we will eventually generate a set with a larger cardinality ($2^\aleph$). The important point is that Cantor claims we can do something to an infinite set to make it bigger, which does not jive well with our traditional understanding of infinity. In fact, this opens up the paradoxical situation of ever-expanding infinite sets, which means there is no natural end point to provide a foundation for set theory.

Addition	Power Set (2^n)	Multiplication
0 + 1	1	0 + (1) - 0
1 + 1	2	1 + (2) - 1
2 + 2	4	2 + (2 x 2) - 2
3 + 5	8	3 + (2 x 2 x 2) - 3
4 + 12	16	4 + (2 x 2 x 2 x 2) - 4
5 + 27	32	5 + (2 x 2 x 2 x 2 x 2) - 5
6 + 58	64	6 + (2 x 2 x 2 x 2 x 2 x 2) - 6
7 + 121	128	7 + (2 x 2 x 2 x 2 x 2 x 2 x 2) - 7
8 + 248	256	8 + (2 x 2 x 2 x 2 x 2 x 2 x 2 x 2) - 8
...	...	...

For the sake of argument, we will assume that it makes sense to say actual infinity exists and that we can generate the power set of an infinite set. What does this have to do with the argument about infinity? As it turns out, Cantor argues that the size or cardinality of the real numbers equals the power set of the natural numbers, which is why the real numbers are not countable. Just as we cannot use the natural numbers 1, 2, and 3 to count the 8 sets of the power set of 3 (see below), because we do not have enough numbers to count the other 5 members of the power set (no bijection), Cantor claims the same goes for the natural numbers and the real numbers. His argument works only if it makes sense to talk about a set of natural numbers reaching completion (the Axiom of Infinity), such that $\aleph + 1 = \aleph$. Given that the natural numbers are defined by our ability to always add more numbers (n + 1), we reach a contradiction if this process reaches completion. If adding and multiplying numbers does not make an infinite set larger, why would taking the exponent of a set (an arithmetic operation) make it larger?

Set		Power Set
2	→	(2)
4	→	(4)
7	→	(7)
		(2, 4)
		(2, 7)
		(4, 7)
		(2, 4, 7)
		(0)

This raises two problems. First, we seem to be talking about two different things. The first column includes numbers and the second column includes sets of numbers, which could not be used to solve problems. In other words, the real numbers and the power set of the natural numbers might have the same size or cardinality, but they do not consist of similar elements. That is, the power set of the natural numbers does not generate the set of real numbers. The most important point is that this step hinges on Cantor's claim that we can think of the set of natural numbers as a complete set, like (2, 4, 7), which allows him to claim there is no way to bridge the gap with addition or multiplication between the set of natural numbers and the power set of the natural numbers. We are left with the idea of a quantity of objects that cannot be counted—what does this mean?

Second, the same logic that Cantor applies to expanding the size of the natural numbers apparently does not apply to contracting the size of the natural numbers. For example, consider the following display of generating the even numbers and the power set from the natural numbers.

Even		Natural		Power Set
2	←	1	→	(1)
4	←	2	→	(2)
		3	→	(3)
		4	→	(4)
				(1, 2)
				(1, 3)
				(1, 4)
				(2, 3)
				(2, 4)
				(3, 4)
				(1, 2, 3)
				(1, 2, 4)
				(1, 3, 4)
				(2, 3, 4)
				(0)
				(1, 2, 3, 4)

The question becomes, in the name of symmetry, if it makes sense to say the power set of the natural numbers, which is constructed from the natural numbers, is larger than the natural numbers, why does it not make sense to say the set of even numbers, which is also constructed from the natural numbers, is smaller than the natural numbers? However, Cantor claims the size or cardinality of the even numbers and the size or cardinality of the natural numbers is the same, even though the two sets clearly differ by a factor of two as the process goes to infinity. Cantor's response would be to invoke his one-to-one correspondence or isomorph methodology. However, this methodology invalidates itself because although we can start the process of pairing up the natural numbers and the real numbers, Cantor claims they are not the same size. Just as we can pair up the even numbers and the natural numbers ad infinitum, without knowing the result, we can pair up the natural numbers and the power set of the natural numbers ad infinitum, without knowing the result.

But again, let us suppose for the sake of argument that the power set of the natural numbers really cannot be counted by the natural numbers, we still have to show why the cardinality of the real numbers equals the power set of the natural numbers. To do this, Cantor invites us to imagine a matrix of infinite binary decimals (1s and 0s), where each row represents a unique subset of the natural numbers. Given that the cardinality of the set of numbers in an infinite decimal is ℵ, we can use 1s and 0s in binary decimals to construct the power set of the natural numbers (2 x 2 x 2 x ... x 2 = $2^{ℵ}$).

.1	1	1	1	1	1	1	1	1	...
.1	0	1	0	1	0	1	0	1	...
.0	1	0	1	0	1	0	1	0	...
.1	0	0	0	0	0	0	0	0	...
.0	1	0	0	0	0	0	0	0	...
.0	0	1	0	0	0	0	0	0	...
.0	0	0	1	0	0	0	0	0	...
.0	0	0	0	1	0	0	0	0	...

According to this methodology (this is admittedly difficult), if a box has a 1, the corresponding natural number is included; if a box has a 0, the corresponding natural number is not included. For example, the first row has a 1 in every box, which means the first row represents the natural numbers (1, 2, 3, ...). The second row has a 1 in every other box, beginning with the first box, so this row represents the odd numbers (1, 3, 5, ...). The third row has a 1 in every other box, beginning with the second box, so this row represents the even numbers (2, 4, 6, ...). The fourth row has a 1 only in the first box and represents 1. The fifth row has a 1 only in the second box and represents 2, and so on for each possible subset.

As we work our way down, we can continue to add more subsets until we arrive at the set of all subsets of the natural numbers (the power set). Most important, Cantor claims that every real number can be expressed as a binary decimal through a conversion process. Therefore, Cantor claims we will run out of natural numbers before we generate the set of real numbers (the real numbers are not countable) because Cantor argues that the size or cardinality of the set of real numbers is equal to the power set of the natural numbers.

The most important question is why did Cantor opt for this peculiar methodology, which does not seem mathematical? I am not a Cantor scholar but I would surmise that his choice had something to do with his desire to generate the real numbers organically from the natural numbers, just as we can generate the integers and the rational numbers from the natural numbers. The reason for this should be obvious: if Cantor wants to "arithmetize" the continuum, he has to show that the real numbers are an organic extension of the natural numbers. That is, he has to show that the continuum can grow organically from the discrete.

However, given that there is no methodology to generate transcendental numbers from natural numbers via the arithmetic operations of addition, subtraction, multiplication, division, or exponent, there must be a way to expand the natural numbers organically without making a clean break from the natural numbers; otherwise, this would end the goal of "arithmetizing" the continuum. Cantor's first insight was probably the observation that the infinite decimal is the link between the natural numbers and the real numbers. All real numbers can be expressed as infinite decimals and the cardinality of every infinite decimal (the quantity of numbers in the infinite decimal) is ℵ, which is the same cardinality of the natural numbers. Thus, we can put the natural numbers into a one-to-one correspondence or isomorph with the numbers in the infinite binary decimals.

Given Cantor's desire to define the real numbers as being organically derived from the natural numbers, he had to find a methodology that would allow him to prove that it was possible to "arithmetize" the continuum. Therefore, if he needed a methodology to show that any given natural number would or would not be included in the various subsets of the natural numbers, he had to opt for a binary system of 1s and 0s, such that the number in the tenth place (1) was either included or not in the subset, the number in the hundredth place (2) was either included or not in the subset, and so on. Given that each number in a binary decimal is 1 or 0 (2 options) and given that the size or cardinality of every infinite decimal is ℵ, there are $2^{\aleph}$ possible infinite binary decimals, which we know from our previous analysis is the power set of the natural numbers, which is achieved by multiplying 2 by itself an infinite number of times ($2 \times 2 \times 2 \times \ldots \times 2 = 2^{\aleph}$). Therefore, if we can show that the cardinality of the real numbers is $2^{\aleph}$ and that the power set of the natural numbers (in binary decimal form) generates the set of real numbers, then Cantor's proof will be confirmed.

However, as I will show in the next chapter, Cantor's desire to "arithmetize" the continuum apparently blinded him to a simpler way to solve the problem. In short, we have already proven that if we accept Cantor's claim that $\aleph < 2^{\aleph}$, then we generate a contradiction if we also accept his claims that $\aleph \times \aleph = \aleph$ and $\aleph + \aleph = \aleph$. We can use simple arithmetic to prove that the cardinality of the real numbers is not $2^{\aleph}$ and that the cardinality of the rational numbers (as decimals) is not $\aleph$. In fact, I will show that real numbers and the rational numbers (as decimals) have the same cardinality ($10^{\aleph}$) if actual infinity is real, which in turns ends the quest to "arithmetize" the continuum.

Conclusion

The examples of Calculus should give us a lot to think about in terms of dynamic infinity. The first is the recognition that the idea of infinity is a product of a finite human mind striving to understand the world, a framework for thinking about the world. If our minds were incapable of thinking in terms of limits, it might never occur to us to calculate a derivative to determine the slope of a curved line at a particular point, or to calculate an integral to determine the area under a curved line.

The second is the realization of what is on the "other side" of infinity. As Plato noted, we use mathematics to solve problems within the world of perception, and the idea of infinity does not allow us to transcend this world. There are no numbers in the Platonic realm, and we do not require an infinite set of numbers in the Platonic realm to provide a foundation for mathematics. The limit process allows us to gain impressive levels of granularity and allows us think about problems in new ways, but when the ad infinitum process is taken to its natural limit, the Δx's cancel and we open the door to see the same world we see every day.

Thinking about the world in terms of limits and infinite processes sets the stage for solving derivatives and integrals, but all the problems of Calculus are solved with the tools of algebra, trigonometry, and other tools that make no use of infinity. The fact that an ad infinitum process ends in a finite solution does not mean infinity is real or that infinity was required to solve the problem. If we estimate the slope of a curved line with a pencil and ruler to arrive at an estimate of 2, then we know with certainty that the Δx's of the ad infinitum process cancel out because we would never arrive at a slope of 2 if we were to face the usual problems of dividing by 0.

Regarding set theory and the power set, I hope I provided useful insights about Cantor's argument and demonstrated that Cantor's argument does not hold. Even if we grant Cantor his assumptions, no matter how controversial, his argument can still be shown to have flaws. If Cantor presumes that the natural numbers and the even numbers are the same size or cardinality merely because we can begin the process of pairing them up, then the same can be said for the natural numbers and the power set of the natural numbers. A discrete pairing process never reaches actual infinity, which means there is no reason to give the power set special status or attention.

Chapter 7: Counting the Real Numbers

Before continuing, we should take stock of where we are in the analysis and consider what variables will be relevant moving forward. When I say Cantor's argument is flawed, we can consider this claim from two perspectives. First, we can consider the philosophical elements, such as whether actual infinity is real or meaningful or whether it makes sense to talk about the power set of an infinite set (these are not mathematical problems). As should be clear from this book, I do not agree with the idea of actual infinity for mathematics because we can show that it involves a logical contradiction when we impose the points of the real numbers onto a finite continuum. The points of the real numbers either have dimension or they do not. If they lack dimension, then the continuum lacks extension. If they have dimension, then the continuum extends to infinity. There is no way to evenly distribute an infinite set of points along a finite continuum.

Although I consider this simple proof against infinity relevant and true, I understand that it will not convince everyone and that many people will continue to believe that it makes sense to say every finite continuum contains the set of real numbers. Therefore, I will not focus any more attention on this point and instead focus on Cantor's errors within his own assumptions. That is, for the sake of argument, I will grant Cantor's claim that actual infinity is real and that it makes sense to talk about the power set of an infinite set, and then see where it leads. Both sides can talk until they are blue in the face about the philosophy of infinity, and one of the primary complaints from the Cantor camp is that the various attempts to discredit Cantor do not address his proofs. For this reason, I will continue to focus on directly discrediting the mathematics of Cantor's argument, not the philosophy.

Second, therefore, we can consider the mathematical elements of the argument. To repeat, Cantor argues that the real numbers are not countable because the cardinality of the real numbers equals the cardinality of the power set of the natural numbers, and these two sets cannot be put into a one-to-one correspondence or isomorph. To do this, Cantor shows that he can construct the set of real numbers from the set of all subsets of the natural numbers (the power set) in the form of infinite binary decimals. As crazy as this sounds, I will grant Cantor this point as well. The Cantor camp will probably be pleased to hear this and might be tempted to stop reading and claim victory, but this is precisely the point where the argument gets interesting.

To set the stage, consider a scenario in which a man painted black on one side and white on the other side strolls down the middle of a football field. From the perspective of the fans one side, the man looks white. They can take pictures and ask other fans on their side of the field to confirm what they saw. From the perspective of the fans on the other side, the man looks black. They can also take pictures and ask others fans on their side to confirm what they saw. As it turns out, however, both are right and both are wrong at the same time. The same goes for infinity. Even if we grant that Cantor's methodology proves that the cardinality of the real numbers between 0 and 1 is $2^{\aleph}$ if we view the problem through the prism of the power set, if we view the problem from another perspective (the set of infinite base-10, place-value decimals), we can prove with simplicity that the cardinality of the real numbers between 0 and 1 is $10^{\aleph}$. Building on this, we can also prove with simplicity that the cardinality of the rational numbers as decimals between 0 and 1 is also $10^{\aleph}$. If that is not interesting enough, I will also show that it is possible to disentangle the density of the rational numbers and the real numbers and arrange them sequentially to have a one-to-one correspondence or isomorph with the natural numbers, as long as we keep an open mind about the rules of the game.

The most important conclusion from my proof is the insight that the realm of infinity is either not real or impossible to define, and therefore impossible to compare one infinite set to another. If one set of valid assumptions tells us one thing about infinity, such as proving that the cardinality of the real numbers is $2^{\aleph}$, and another set of valid assumptions tells us another thing about infinity, such as proving that the cardinality of the real numbers is $10^{\aleph}$, then the problem lies with infinity, not with the methodologies. On top of this, although this is more difficult to prove, it shows that how we think about infinity is a function of our numbering system. For example, if we consider the rational numbers as fractions, we get one result for the cardinality, but if we consider the rational numbers as decimals, we get another result for the cardinality. In short, I am not asking the Cantor camp to reject Cantor's argument. I am asking them to consider equally valid arguments with different results.

The Diagonal Argument

Cantor's diagonal argument is one of the most controversial mathematical proofs in history. This is unusual because there should never be controversy in mathematics about proofs, at least in theory; something is either true or it is not, and the proper use of mathematics should leave no confusion as to whether the conclusion follows necessarily from the steps of the proof, unless the proof rests on assumptions that are not intuitive or for which there is limited agreement, such as the Axiom of Infinity. To the Cantor camp, Cantor's diagonal argument is clear and irrefutable. To the apostates, however, like myself, the proof demonstrates nothing significant and in many ways supports the argument against actual infinity, for reasons I will explain. It all depends on the lenses you use to view the problem. Because I am in the potential infinity camp, I tried my best to overcome any biases or errors I might have, but the longer I studied it, the more it confirmed my own beliefs about potential infinity. Even after I finally understood the importance of the power set for Cantor's proof, I was still not convinced that Cantor's diagonal argument, by itself, proved what it claimed to prove.

There are two reasons for the confusion. First, as is the case with most issues in this book, the truth or falsity of the diagonal proof boils down to who is right about infinity, potential or actual, which is ultimately a philosophical problem. The diagonal proof does not settle the issue, which means the proof will motivate each side to double down. I would say the proof is structured in an ad infinitum way that suggests potential infinity wins the day, but I can see how the Cantor camp would disagree. Second, the proof is a visual display of Cantor's two most important assumptions: actual infinity and the power set of the natural numbers. If you begin with Cantor's argument and fully digest what it says, then the diagonal proof makes sense. However, this does not work the other way. We cannot ponder the diagonal argument by itself and reverse engineer the conclusions about actual infinity or power sets. Thus, as we will see, the biggest problem with the diagonal argument is that it proves nothing at face value. The diagonal proof is a visual display of more fundamental claims, and the proof makes sense only if you already understand and agree with those more fundamental claims.

The debate is filled with members of the Cantor camp expressing their frustrations with the apostates, and vice versa. The Cantor camp points to the diagonal number that is generated and the fact that it is not on the list (more on this below), insisting that this proves their point, only to have the apostates agree that the diagonal numbers is not on the list but noting that this proves nothing because the process continues ad infinitum for the natural numbers as well. And if the Cantor camp insists that the argument is valid due to the properties of power sets, the apostates would respond that it does not make sense to take the power set of an infinite set because it does not make sense to talk about infinite sets—and so on, ad infinitum.

With this introduction out of the way we can turn to Cantor's diagonal argument. Of note, Cantor begins by supposing that his opponent's argument is correct, namely, that the real numbers are denumerable or can be put into a one-to-one correspondence or isomorph with the natural numbers. He then proceeds to show that this argument results in a contradiction, which supposedly proves that his argument must be correct.

This style of proof is known as *reductio ad absurdum*. This type of logic works for binary situations, either/or, but even then we have to be careful because defining the either/or options requires precision. For example, in the case of the diagonal argument, does the alleged failure of the argument mean that Cantor's claim is necessarily correct? In other words, even if we accept that the real numbers are not countable, does this necessarily mean that the set of real numbers is larger than the set of natural numbers? After all, perhaps the natural numbers also are not countable? Or, perhaps it is possible that we cannot establish a one-to-one correspondence or isomorph between the natural numbers and the real numbers because they are two different kinds of sets (the infinity of addition and division)? Thus, our first warning sign is that Cantor attempts a *reductio ad absurdum* where it might not apply.

Cantor begins with a matrix similar to the ones we used previously to demonstrate a one-to-one correspondence or isomorph between two sets of numbers, with the natural numbers paired up with a list of random real numbers between 0 and 1 (that extend to actual infinity via "...") in which every real number appears only once:

Natural	Real
1	0.1849642...
2	0.9568235...
3	0.5710432...
4	0.2389485...
5	0.7015294...
6	0.0691428...
7	0.3541389...
...	...

Cantor's most important claim is that no matter how long we make this list, there will always be at least one real number that is not on the list. To support his claim, he uses the diagonal method to prove that every list of real numbers will necessarily be incomplete. If we accept that each real number on the list is unique, then we can create a new number that is not on the list by taking the bold numbers that form a diagonal and change at least one of them to generate a new number. For example, in the first real number, 0.1849642..., if we change only the first number, such as making it a 2, then we will not repeat this real number because each real number is listed only once. Therefore, if we take the diagonal numbers in bold, 0.1519229..., and change at least one of the numbers, we are guaranteed to generate a number that is not on the list (up to that point), although we should keep in mind that this new diagonal number is a finite rational number every step of the way. Thus, Cantor concludes that the real numbers are therefore not countable, which contradicts the original assumption that they are countable.

If you are like me, you are probably looking at this argument and scratching your head—so what? At first glance, it seems to prove something important (that every finite list of real numbers is incomplete), but that applies to any list of numbers. The process of building a list of real numbers one real number at a time will never reach completion. That is, there is nothing in this diagram to suggest the idea that the real numbers are generated from the natural numbers or that the cardinality of the real numbers equals the power set of the natural numbers, which in turns means that there are not enough natural numbers to count all the real numbers. With all of this said, we should now turn our attention to the weaknesses of this argument.

First, the argument claims that any finite list of real numbers will necessarily be incomplete. This is true, but I cannot imagine anyone disagrees with this, even people who are members of the potential infinity camp. Cantor states that if we tweak the numbers along the diagonal, we can generate a new number that is not on the list up to that point. However, the problem with this argument is the same logic applies to the natural numbers in the left column. No matter how long we make the list, we can always add 1 to the last natural number to generate a new natural number not currently on the list. For example, the current list ends at 7, but there is nothing to stop us from adding 8 to the list, and so on. Regardless of what Cantor says about the power set, we can add 1 to any natural number and the process will never reach completion, according to the potential infinity perspective. If anything, Cantor's argument shows that the idea of an infinite set of numbers is self-contradictory because a discrete process of counting individual numbers never reaches termination.

Second, the diagonal argument claims that the list will be incomplete without explaining why the list will be incomplete. The secret sauce here is that Cantor equates the counting of natural numbers with the counting of numbers in place-value positions of infinite decimals. For example, if our list were to stop at the first decimal point (0.0 0.1, 0.2, ..., 0.9), then if we were to select any one of those 10 numbers, our list would be incomplete because we would miss the other 9 numbers. Likewise with the next step, although the natural numbers transition from 1 to 2, the diagonal number transitions from 10 possibilities to 100 possibilities. If we consider two-digit decimal numbers, there are 100 possibilities (0.00, 0.01, 0.02, ... , 0.99). Therefore, if our diagonal two-digit number is one particular number, then we have missed the other 99 possibilities. If we consider the seven-digit number from the above matrix, 0.1519229..., then if we want to create a new number, we should keep in mind that 0.1519229... is one of 10,000,000 possible numbers. Therefore, we should not be surprised when we are able to create a new 7-digit number that is not on the list because our diagonal number is only one of 10 million possibilities. What we can say with certainty is that the real numbers grow by a factor of 10 relative to the natural numbers with each decimal place.

If we want to avoid this situation (generating a number that is not on the list), all we have to do is make sure we include every possible number before moving on to the next place-value number. That is, if we list all the 10 single-digit decimals (0.0, 0.1, 0.2, ... , 0.9) before we list all the 100 double-digit decimals (0.00, 0.01, 0.02, ..., 0.99), and so on, then we avoid a scenario in which any diagonal numbers are missed along the way, and will generate every possible real number as x goes to infinity. For example, if we listed all the 10,000,000 7-digit decimals from the above matrix, then Cantor would be wrong to say we will have missed at least one number.

Cantor would probably claim that this process will never generate irrational numbers (infinite decimals that never terminate or repeat) but this merely reinforces our claim that discretely counting numbers ad infinitum will never reach actual infinity, which is true whether we are talking about natural numbers or real numbers. Thus, when we consider Cantor's diagonal argument, the point is not whether he is right or wrong about the relative sizes of the natural numbers and the real numbers, but the recognition that the diagonal argument itself does not prove anything definitive. The diagonal argument proves nothing about infinity and gives us no insights about infinity.

The above points suffice to prove that the diagonal argument as a stand-alone argument does not prove what it claims to prove. It is a visual display with invisible assumptions. To remove all doubt about the failure of the argument, we can consider it from a different perspective—the dreaded switch from counter-clockwise to clockwise driving. That is, rather than begin with a list of random real numbers we will begin with a list of random rational numbers, keeping in mind that repeating rational numbers and irrational numbers both extend to infinity, such that the cardinality of the set of 3s in 1/3 = .333... has the same as the cardinality of the set of numbers after the decimal for any transcendental number. And keeping in mind that 0 is a number, even terminating rational numbers can be thought of as extending to infinity.

Natural	Rational
1	0.2500000...
2	0.3333000...
3	0.1666666...
4	0.1250000...
5	0.0123123...
6	0.7521500...
7	0.5000000...
...	...

Needless to say, the same argument applies, which means Cantor is in the awkward position of either rejecting his claim that the diagonal argument proves his point or admitting that the rational numbers are not countable. Using the same diagonal principle, we can say with confidence that every finite list of rational numbers is incomplete, just as every finite list of natural numbers is incomplete. In this case, we can tweak the diagonal number 0.2360100 and change at least one number to show that the new diagonal number (a finite rational number) is not on the list, and the process continues ad infinitum.

Cardinality Redux

When I previously attempted to count the rational numbers with fractions, we saw that the process generated duplicate entries and that we could mentally delete these duplicate entries (1/2 is replaced by 2/4, 1/3 is replaced by 2/6, etc.) to prove that the cardinality of the rational numbers generated with fraction between 0 and 1 is ℵ.

0/ℵ, 1/ℵ, 2/ℵ, 3/ℵ, ... , ℵ/ℵ

Of interest, Cantor believed he proved that the cardinality of the entire set of rational numbers (not only between 0 and 1) is ℵ, via his one-to-one correspondence or isomorph diagram. However, inserting numbers into a diagram is not the same as using mathematics to calculate the size or cardinality of a set because a diagram might not take into account all the invisible variables or assumptions. If we accept my proof that the cardinality of the rational numbers generated from fractions between 0 and 1 is $\aleph$, then if we recall that the cardinality of the natural numbers is $\aleph$, the cardinality of the entire set of rational numbers is $\aleph \times \aleph \times \aleph \times \ldots \times \aleph = \aleph^{\aleph}$, which is much larger than the cardinality of the real numbers ($2^{\aleph}$), per Cantor's calculation. Although Cantor claims that $\aleph \times \aleph = \aleph$, not $\aleph \times \aleph = \aleph^2$, I previously proved that this generates a contradiction with Cantor's claim that $\aleph < 2^{\aleph}$ because it means that $\aleph < 2$. Thus, either Cantor's proof of the countability of the rational numbers is incorrect or different methodologies generate different results, which means set theory is inconsistent.

Keeping in mind our image of the man painted half white and half black walking down the middle of a football field, let us consider another perspective. If we want to get an accurate count of the size or cardinality of the rational numbers, we could employ a different methodology to eliminate the duplicate entries. Consider the following:

$1/2 = 2/4 = 3/6 = 4/8 \ldots = 0.5$

$1/4 = 2/8 = 3/12 = 4/16 \ldots = 0.25$

$1/8 = 2/16 = 3/24 = 4/32 \ldots = 0.125$

It should be clear that there is an endless list of different fractions that equal 0.5, an endless list of different fractions that equal 0.25, an endless list of different fractions that equal 0.125, and so on. What these rational number fractions have in common is that they can be represented by a unique decimal such that 0.5 represents the set of all the rational number as fractions that equal one half, and so on. Therefore, we can think about the rational numbers as decimals and see what happens when we calculate the size or cardinality. We will still get duplicates (0.5, 0.50, 0.500, ...) but the decimal format will make it easier to calculate the cardinality of the rational numbers.

To keep things simple, we will construct our set of rational number decimals one place-value decimal at a time. For example, if we begin with the first place-value after the decimal, the size or cardinality of this set of rational numbers is 10 (0.0, 0.1, 0.2, 0.3, ...0.9). If we use the first two place-values after the decimal, the cardinality of this set of rational numbers is 100 (0.00, 0.01, 0.02, 0.03, ... , 0.99). If we use the first three place-values after the decimal, the cardinality of this set of rational numbers is 1,000 (0.000, 0.001, 0.002, 0.003, ... , 0.999), and so on. Thus, if we want to generalize this process, keeping in mind that every decimal along the way is a rational number, then the cardinality of the set of rational numbers at any point during this process is 10^n, where n represents the quantity of place-value positions to the right of the decimal.

The question now becomes, how long can this process continue? There is no theoretical limit to how far we can go. Each step of the way we will have decimals that represent rational numbers. However, in the context of infinity, if we consider that each place-value number is a natural number, then the natural numbers in the decimals are countable, according to Cantor. This means the cardinality of the set of natural numbers for each rational decimal as n goes to infinity is $\aleph$, which allows this process to generate 1/3 = .333... and every other possible combination of numbers that extend to infinity. Therefore, if we continue this process as n goes to $\aleph$, then the cardinality of the rational numbers between 0 and 1 is $10^{\aleph}$, which is different from our previous calculation of $\aleph$ when we used fractions, which raises questions about why rational numbers as fractions and rational numbers as decimals generate different results. And if we consider that this process gets repeated between every two natural numbers, then the cardinality of the entire set of rational numbers as decimals is $(10^{\aleph})^{\aleph}$.

I have no doubt that this proof will not convince the Cantor camp, but my proof relies on basic arithmetic and the same assumptions that Cantor relies on for his proofs, such as the infinite binary decimals used for the power set. If the objection is that the one-to-one correspondence or isomorph methodology proves that the natural numbers and the rational numbers have the same cardinality, I would counter that I have already raised valid concerns about this methodology. If the objection is that the decimals of rational numbers cannot extend to $\aleph$, the Cantor camp will have to explain when the process ends, or when we make the transition from the finite to the infinite. Does the set of 3s in 0.333... have a cardinality of $\aleph$ or not? Do transcendental numbers extend beyond this point?

Building on this, we can use a similar methodology to show that the rational numbers and the real numbers have the same cardinality. If we consider that the real numbers between 0 and 1 represent the set of infinite decimals, then if each decimal place contains 1 of 10 possible natural numbers and the cardinality of the set of natural numbers for each infinite decimal is $\aleph$, then the cardinality of the real numbers is $10^{\aleph}$, which required nothing more than basic arithmetic to prove. My sense is that Cantor's desire to "arithmetize" the continuum by making the real numbers an organic extension of the natural numbers via the power set and infinite binary decimals might have blinded him to this obvious truth. If we need to explain how the rational numbers as decimals and the real numbers could possibly have the same cardinality, we could suggest that the two sets of numbers blend into the same set of numbers at $\aleph$, which is the point where adding more numbers does not change the value of the number, such that $\aleph + 1 = \aleph$. Of interest, we can take the power set of a set of numbers but not of a set of numbers in an infinite decimal (to make the decimal longer), which is an Achilles heel for Cantor. As I will address, the distinction between the rational numbers and the real numbers is the distinction between the infinity of addition and the infinity of division.

With this, we now understand why Cantor thought the cardinality of the real numbers was $2^{\aleph}$. It had nothing to do with the cardinality of the real numbers being the power set of the natural numbers and everything to do with the cardinality of the infinite binary decimals. If each number in an infinite decimal is 0 or 1 (2 options) and the cardinality of the set of natural numbers for each infinite binary decimal is $\aleph$, then the cardinality of the infinite binary decimals is $2^{\aleph}$. However, given that the real numbers have 10 options for each decimal place, not 2, Cantor's argument was wrong by a factor of 5. It is not clear what is to be gained by trying to express infinite decimals as infinite binary decimals, aside from a lot of extra work and confusion.

The really bad news for the Cantor camp is that if my solution is correct and the set of real numbers cannot be generated from the set of natural numbers via the power set, then we have proven that the real numbers cannot be derived organically from the natural numbers, which means the goal to "arithmetize" the continuum has failed. Just as bad for the Cantor camp, given that different yet valid methodologies for calculating the cardinality for sets of numbers produce different results, we can conclude that the problem lies with the idea of infinity, not with the mathematics. As I said before, we are left with the conclusion that either infinite sets do not exist or they are all the same size, despite different notations. Even worse, if different methodologies produce different results in set theory (if we can show that the claim that cardinality of the rational numbers equals ℵ is both true and false), then this proves that set theory is inconsistent.

Symmetry

If we agree that discrete counting never reaches actual infinity, then we should take a different approach, such as thinking about actual infinity within our field of vision to explain what it looks like. For example, we cannot conceptualize or imagine an actual infinity of numbers, but we can imagine a continuum from 0 to 1 and fill in some of the blanks to "see" it in ways that we could not before.

[0.0 0.1 0.2 0.3 0.4 0.5 0.6 0.7 0.8 0.9 1.0]

We can agree or disagree on the possibility of infinity, but the important point is that we can see the numbers between the brackets and at least imagine it filled with infinitely many points. As a mental exercise, I would propose that we push the symbolic nature of mathematics to the limit and think about the decimals between 0 and 1 as nothing more than infinite sequences of symbols, devoid of any reference to anything in the world. In fact, rather than use numbers, we could think of the symbols as representing weights, such that .01 is ten times heavier than .1, .001 it ten times heavier than .01, .0001 is ten times heavier than .001, and so on, which is the opposite of how we normally think of numbers—smaller is bigger. If this difficult to understand, imagine a 10-pound weight hanging from a horizontal pole affixed to a wall, such that the farther the 10-pound weight is from the wall, the more torque the weight puts on the pole.

```
[==============================================O
```

10 lbs	10 lbs	10 lbs	10 lbs	Individual Weights
	100 lbs	1,000 lbs	10,000 lbs	Total Torque

If the wall is on the left, then if the first 10-pound weight is one unit away from the wall (one decimal place), it will exert 10 pounds of torque. If the weight is two units away (two decimal places), the torque will increase tenfold, and so on. What this means is that we will weigh numbers based not on their mathematic value but on their weight, such that $.1 < .01 < .001 < .0001$, etc.

What this means is that the "lightest" decimals between 0 and 1 are .1, .2, .3, and so on. After we reach .9, the next lightest number is .01, followed by .02, .03, and so on, all the way to .99, which is followed by .001, and so on. Likewise, on the other end of the continuum, .999... will be the heaviest number preceded by .899..., .799..., and so on.

This discrete counting process (from small to large or from large to small) will never result in every decimal being listed because a discrete counting process never reaches infinity, whether we are talking about natural numbers or real numbers. For this reason, rather than attempt to count the real numbers, we will focus on their weights. As our mental exercise continues, we can imagine a scenario in which the real numbers between 0 and 1 are suspended in a special fluid such that we can shake them up and allow the real numbers to settle down according to their weight, without having to rearrange them one by one. If we assume that every decimal moves to its appropriate place in a split second, such that 0.1 is first and 0.999... is last, then we can imagine the real numbers reorganizing themselves in the blink of an eye. Just as we can pour three fluids with three different densities or specific gravities into a glass and watch the three fluids reconstitute into three visible layers, if the weights of the real numbers cause them to move to their correct place in a snap without error, such that .997 slides in between .996 and .998, then we have resolved the density problem by making the real numbers sequential in a special way. We might have to tap the container of fluid a few times to make sure none of the real numbers stick, but if we imagine a sensitive fluid that distinguishes every real number with infinite precision, then we will assume success.

With this mental exercise out of the way, we can see the final result.

0.1

0.2

0.3

...

0.14159265358979323846... (π - 3)

...

0.71828182845904523536... (e - 2)

...

0.799...

0.899...

0.999...

The importance of this display (not a list!) of the weighted real numbers, which are finally sequential and no longer burdened by the problem of density, is that we can include irrational numbers without violating the claim of the Cantor camp that a list of real numbers will never reach a number like e - 2 because the number is infinitely long. Discrete counting from .1 will never reach the two irrational numbers in the display, but every irrational number has an infinitely precise position in the display. The reason for this is every real number is greater or less than every other real number, which means every real number must have a relative position to every other real number and therefore must have an infinitely precise position in this display. That is, if we are going to claim e - 2 is a real number, then there must be a way to derive it with infinite precision, and it must differ from every other real number such that it has a unique position in the display. The same goes for every real number between 0 and 1.

With that admittedly painful mental exercise out of the way, we are now ready to reveal why it matters. The next step, the shift from counter-clockwise to clockwise driving, is to take this visual display of the weighted real numbers between 0 and 1 and take the mirror image with the decimal point as the hinge.

1.0	0.1
2.0	0.2
3.0	0.3
...	...
...64832397985356295141.0	0.14159265358979323846...
...	...
...63532540954828182817.0	0.71828182845904523536...
...	...
...997.0	0.799...
...998.0	0.899...
...999.0	0.999...

At first glance, it is not clear what we are looking at on the left side, but upon closer inspection it should be clear that we are looking at a display (not a list!) of the natural numbers that we discussed previously, from 1 to ...999, the smallest and largest natural numbers, respectively. We cannot make a complete list of the natural numbers, just as we cannot make a complete list of the real numbers between 0 and 1, but we know that every natural number has a specific place in the set of natural numbers, such that each natural number is less than or greater than every other natural number. As surprising as this sounds, this means we have established a one-to-one correspondence or isomorph between the real numbers and natural numbers, contradicting Cantor's claim in the diagonal argument. If the Cantor camp balks at this claim, they now have a better idea of why some people question their idea of actual infinity.

The validity of this display hinges on the idea that it makes sense to talk about the largest natural number, ...999, which is one greater than ...998. In fact, if we consider that the mirror image of every transcendental number has a precise position in this display of the natural numbers, just as every transcendental number has a precise place in the display of the real numbers, it means the transcendental numbers are a subset of the natural numbers. The good news of this conclusion for the Cantor camp is that it confirms Cantor's definition of an infinite set because we have now shown that the real numbers can indeed be put into a one-to-one correspondence or isomorph with a proper subset if itself (the natural numbers).

To be clear, I do not accept this analysis or this mirror-image display because I do not believe in actual infinity. I also do not agree with the idea of the largest natural number, but the Cantor camp has to take it seriously if they expect us to take them seriously about infinite sets and infinite decimals. This analysis merely shows the logical conclusion of belief in infinite sets. The purpose of this mental exercise was to demonstrate the paradoxes and contradictions that arise when we assume that actual infinity is a real, as well as to show that we can talk about putting the natural numbers and the real numbers into a one-to-one correspondence or isomorph if we think about it in creative ways. The real numbers and the rational numbers are dense and therefore not sequential, but our weighted number approach disentangles the density of the real numbers to reorder them in a sequential way.

Normally, the process works the other way. Cantor would make a visual display for a set of numbers (even numbers, integers, rational numbers, and so on) and then show that the set of numbers can or cannot be put into a one-to-one correspondence or isomorph with the natural numbers. In this case, we began by showing that the real numbers can be put into a one-to-one correspondence or isomorph with the natural numbers and will now show what a visual display of the real numbers looks like. In the case of the rational numbers, Cantor made a visual display to show the set of all fractions, but in this case we will make a visual display of the set of infinite decimals. However, given that the rational numbers and the real numbers can both be expressed as infinite decimals, this display will show why the two sets are the same size.

	1	2	3	4	5	6	7	...	n
.1	.1	.2	.3	.4	.5	.6	.7	...	.9
.01	.01	.02	.03	.04	.05	.06	.07	...	.99
.001	.001	.002	.003	.004	.005	.006	.007	...	.999
.0001	.0001	.0002	.0003	.0004	.0005	.0006	.0007	...	.9999
...	...	...	...	...	...	...	...	...	...
$1/10^{n}$	$1/10^{-n}$	$2/10^{-n}$	$3/10^{-n}$	$4/10^{-n}$	$5/10^{-n}$	$6/10^{-n}$	$7/10^{-n}$	...*	.999...

Rather than use the bold perimeter numbers to generate fractions, as in the case of the display for the rational numbers, this display multiplies the bold perimeter numbers to generate the set of infinite decimals. For example, to generate the .5 in the first row, we multiply .1 x 5. In response to probable criticism from the Cantor camp, if it makes sense to use a display to complete the infinite set of natural numbers, we can also use a display to complete the infinite set of infinite decimals. I should highlight two things about this display. First, each row is not the same length. The first row has 9 numbers, the second row has 99 numbers, the third row has 999 numbers, and so on, with the last row representing the process as x goes to infinity. (The fact that the rows are not the same length has no impact on the process of diagonal counting, which Cantor used to show that two sets of numbers were the same size.) Second, every irrational number is located inside the field with the * because every irrational number is infinitely long but less than .999..., the number on the bottom right.

In other words, what this display shows is that if we apply the idea of limits and actual infinity to the rational numbers as decimals (not as fractions), the result is the set of real numbers because this matrix will generate every possible decimal, to include infinite decimals, as x goes to infinity. In other words, the only way to bridge the gap between the infinity of addition and the infinity of division is the transition from potential infinity to actual infinity with infinite decimals. And therein lies the rub: if infinity is not real, then Aristotle was right; but if infinity is real, then the rational numbers as decimals and the real numbers are the same, which is an unwelcome finding for the Cantor camp.

Conclusion

My initial goal when writing this book was to discredit Cantor's claim that the real numbers are not countable by discrediting the idea of actual infinity. I disagreed with his premises that infinite sets were possible and that we could generate the power set of an infinite set, and thought I could provide a reasonable argument to support my claim. However, I realized that these arguments would probably not work with the Cantor camp because they accept the Axiom of Infinity and the Axiom of Power Set. I therefore decided to focus my efforts on attacking Cantor's argument head on by proving it produces inconsistent results. This turned out to be a better approach because I was able to challenge Cantor by granting him his own assumptions. That is, rather than debate Cantor's claim that the cardinality of the real numbers between 0 and 1 is $2^{\aleph}$, I used a simpler method to demonstrate that the correct answer is $10^{\aleph}$, even if we accept Cantor's assumptions about actual infinity.

The weakest link of the actual infinity perspective is the realization that the set of numbers in an infinite decimal has a cardinality of $\aleph$, which means the numbers in infinite decimals are countable. If the numbers of an infinite decimal are countable, it means we can generate every infinite decimal one decimal place at a time as x goes to infinity, whether rational or irrational. This is why the iterative process of generating the set of rational numbers one place-value decimal at a time generates the set of real numbers as x goes to infinity. The set of numbers in the rational number 0.333... and the set of numbers in the irrational number e both have the same cardinality. Therefore, if it makes sense to say 1/3 = 0.333..., then it makes sense to talk about generating irrational decimals with the same iterative process. I have no doubt that the Cantor camp will grumble about this, but the irony is that if they deny that an iterative process is capable of generating an infinite decimal, then the one-to-one correspondence or isomorph methodology fails as well.

Chapter 8: The End of Infinity

One of the most curious ideas I encountered while researching this book was the claim that a finite human mind cannot grasp actual infinity (our intuitions are too limited and are often an obstacle to understanding infinity), and yet at the same time the human mind is capable of developing a valid theory of infinity and discerning which points are correct or incorrect. Cantor made it look like a discovery that took us all by surprise, as if the human mind or our human numbering system played no role in shaping the result. However, rather than think about infinity as something independent of us that we discover, we should also factor in the role of the human mind and our human numbering system in the process and understand that a human mind that is incapable of grasping infinity could not assess which theory of infinity is correct.

Kant argued that the way we perceive the world is a function of the a priori forms of the intellect. In other words, the reason we perceive the world the way we do, in terms of space, time, and causality, is because that is how our mind presents the world to us as our minds transform sense data into perceptions, which includes the continuums of lines, space, and time, which we understand intuitively. Just as a person who wears red sunglasses will see the world as red, the way we perceive the world is in part due to our hardware. This hardware defines the limits of our understanding, and the study of these limitations is the goal of critical philosophy. The most important point is that if Kant was right, then it is impossible to transcend our own limitations to understand the nature of the reality that makes our world of perception possible. We can perceive the world (the phenomenal world), but we cannot speculate on the world behind the world of perception (the noumenal world), if it even makes sense to talk about a world behind the world of perception. What this means is that if our hardware is designed in such a way that we cannot grasp infinity (if our intuitions are incapable of grasping infinity), then we cannot make a valid theory about infinity. However, if Kant was wrong and we are capable of making a valid theory about infinity, then we are capable of understanding it.

In other words, what my research revealed was that what really matters when we talk about infinity is the metaphysics of infinity, not the mathematics of infinity. The mathematics of infinity makes philosophical assumptions about infinity, and most of the confusion stems from metaphysics creeping into mathematics. We know that Cantor was a religious man and was convinced that his theory of infinity had metaphysical and religious implications (he would spend his latter years in contact with Catholic priests discussing the religious implications of his theory), so in this final chapter I would like to transition to the metaphysics of infinity to more closely consider the philosophical implications of infinity.

Embodied Mathematics

One of the themes I have raised in this book is the idea that the numbering system we use reflects the strengths and weaknesses of the human mind, that the base-10, place-value numbering system is just one of many possible numbering systems, and that the numbering system we use bakes some interesting things into the cake, just as the language and grammar we use influences how we think. What this means is that we should consider the philosophy of mathematics and infinity in the context of the human mind that performs the mathematics, but this is one critical piece that seems to be missing from the infinity argument. The argument takes an almost dogmatic approach that seems to ignore the role of the human mind, as if mathematical truths are out there to be learned through a process of objective discovery, with limited interest or concern for the human mind that is searching for and shaping these truths.

One of the other themes I have raised in this book, and something I will develop later in this chapter, is the idea that something other than mathematical thought provides the foundation for mathematical thought. The process of conceptualizing 1 is not grounded in some sort of otherworldly Form or Idea of 1 that we meditate or reflect on. People were thinking and communicating in quantitative terms long before we had a formal numbering system, just as we know that babies are capable of quantitative thought before learning mathematics. Therefore, we should consider the more distal or fundamental modes of thought that shape mathematical thinking. To do this, we will consider the insightful paper by Rafael E. Núñez, *Creating Mathematical Infinities: Metaphor, Blending, and the Beauty of Transfinite Cardinals.*

Núñez considers two main dimensions of human cognitive phenomena. The first of these is what he calls aspect, which characterizes the structure of event concepts, to include iterative actions, such as tapping your finger, and continuous actions, such as a rock falling. These two aspects are common in nature and the mind considers them differently. These events might have a start and an end, such as jumping; might have a start only, such as leaving a place; or might have an end only, such as arriving at a place; and when they have an end, we refer to it as a resultant state.

The most important distinction for aspect is whether it is perfective or imperfective. Perfective aspect has an inherent completion, such as jumping, whereas imperfective aspect has no inherent completion, such as flying (not to be confused with taking off or landing). If we go back to the pre-Socratics, we might use terms like apeiron (unlimited) and peras (limited) to understand the difference between imperfective and perfective. We see the same idea in the grammar of Romance languages, like Spanish, which has one past tense (preterite) for actions that were completed and another past tense (imperfect) for actions that were ongoing in the past. Imperfective can be iterative (such as multiple study sessions before a test) and continuous (such as the idea of sleeping and sleeping and sleeping). The capacity to conceive of the continuous in iterative terms is important for infinity because the idea of potential infinity is based on the idea that something may or may not have a starting point but has no end point—no completion and no final resultant state. Thus, Núñez's key point is that potential infinity (the infinity of Aristotle) has imperfect aspect. Numbers have no completion or final resultant state.

The second main dimension of human cognitive phenomena is what Núñez calls the Basic Mapping of Infinity (BMI), which is seen inside and outside mathematics. The first BMI input space is called the Completed Iterative Processes (CIP) with perfective aspect, which is the realm of the finite. An example of this would be counting the five fingers on one of your hands. This is an iterative process that reaches completion. The second BMI input space is called the Endless Iterative Processes (EIP) with imperfective aspect, which is the realm of potential infinity. An example of this would be counting the natural numbers ad infinitum because it is an iterative process that never reaches completion. According to Núñez, these are the only two direct input spaces for quantitative thinking, with no space for directly conceiving of actual infinity.

To make the transition to thinking about actual infinity, we have to blend these two input spaces, which is done by leaving out the distinguishing elements of the two input spaces. When we do this, however, there is a conflict between one process with an end and a resultant state and another process being endless with no final resultant state, which is why there is so much disagreement about actual infinity. From CIP, two ideas, "process must have an end" and "final resultant state," are projected onto the blended space, ignoring the idea of "finite." From EIP, the idea of "process has no end" is projected onto the blended space, ignoring the idea of "does not have a final resultant state." The blended space is an "endless process" with an "end" and a "final resultant state." In the case of static infinity, this is more difficult to grasp because things like the natural numbers by definition have no end or final resultant state. In the case of dynamic infinity, this is easier to grasp because we can imagine a process that can be thought of as infinitely divisible that has a final resultant state, such as a triangle transforming into a circle. However, at least in the case of Calculus, we can show that dynamic problems can be solved with algebra and that the thought of an infinitely divisible process is not required.

As we addressed previously, one of Cantor's primary objectives was to "arithmetize" the continuum, which means he had to bridge the gap between the numbering system for the finite and the infinite, which means he had to bridge the gap between the infinity of addition and the infinity of division. For this reason, Cantor redefined the natural numbers in a way that would make this possible (we cannot bridge the gap with the traditional definition of numbers), which we addressed with the idea of the sizes of sets or cardinality. To do this, Cantor invoked another metaphor, namely, the idea that the "same number as" is the same thing as being "pairable" via the one-to-one correspondence or isomorph. The "same size" part of the metaphor is that after we have established a one-to-one correspondence or isomorph, such as the five fingers on the right hand the five finger on the left hand, that there is nothing left over, no fingers that have not been counted. This differs from the idea of "more than" in the sense that if one set has more than another set, there is something "left over," such as a person with six fingers. Cantor simply does away with the "left over" part, such as with the natural numbers and the even numbers. The critical and controversial step is that Cantor equated the idea of "as many as" with the idea of being "pairable." Núñez concludes: "Cantor also intended pairability to be a *literal generalization* of the very idea of number. An *extension* of our ordinary notion of 'same number as' from finite to infinite sets. There Cantor was mistaken."

Despite pointing out the obvious errors in Cantor's thought, Núñez is generous to Cantor and gives him credit for creating a new type of mathematics, but at the end of the day Cantor's blended space is at best true metaphorically and only partially. Metaphor is powerful for thinking about the truth (many grand theories are metaphors), but Cantor would undoubtedly deny that his theory of infinity was a mere metaphor. Cantor might argue that Núñez was correct to highlight the limits of the human mind when it comes to thinking about infinity (the two input spaces of the human mind), but he would nonetheless argue that infinity is real and outside the human mind. However, rather than consider Cantor's theory of infinity an intellectual leap, we could also consider it as an attempt to impose the finite onto the potentially infinite. Cantor and the Cantor camp are fond of pointing out the limitations of the human mind, but we could also accuse Cantor and the Cantor camp of being trapped by the finite and not fully understanding of the idea of potential infinity.

The Uncanny Correlation

Regardless of what we think about infinity, there should be no doubt that the correlation between mathematics and the empirical world has an almost mystical quality. How is it possible for an ivory tower mathematician to develop abstract formulas that "work" years or decades later when the science "catches up"? When a formula like $E = mc^2$ is proposed, we have to ask whether this formula merely describes what is happening with a high degree of accuracy or whether it is an algorithm that runs the universe.

For example, if an expert studies a weather app that makes predictions for the upcoming week, he could study the predictions over time and develop his own algorithm to predict what the app will predict. If the expert's model eventually predicts what the weather app predicts with absolute precision, every time, then we can safely conclude that the app and the expert are probably using the same algorithm. This does not speak to the accuracy of the app or whether the algorithm is the actual code for how weather unfolds, but it sets the stage for our discussion. We know that scientists like Newton believed they were discovering the formulas or algorithms that run the universe, so we should consider the implications of this view.

If we begin on this side of reality, where our mathematical formulas merely describe what is happening with varying degrees of accuracy, then we should consider how this is possible. As we addressed before, this bottom-up approach applies to language and mathematics. In the case of language, as we observe the world and ourselves, we use our mouths to articulate arbitrary sounds to understand the world and to communicate with others. At some point, we developed an alphabet to represent these arbitrary sounds, which is a powerful tool that allows us to store our ideas and communicate over distance and time. No human language could ever account for all of reality with a level of precision or granularity down to the level of individual atoms or beyond (an infinite vocabulary), but this is due to the limitations of the human mind and the purpose of language, not to the limitations of language. When we claim there is no theoretical length to a sentence or that we can always add another word to a sentence, we are not at the same time claiming that an infinite sentence actually exists in a Platonic realm. Specialists often have highly refined vocabularies for their fields of study that most people could not grasp, but the real power of language comes from its ability to generalize and streamline our thoughts.

Thus, we are faced with the following situation. On the one hand, there is no doubt that the world consists of unique things or species, like trees and cats, with real scientific distinctions (chemical composition, DNA, etc.). On the other hand, there is also no doubt that language is a reflection of the human mind. For example, conjunctions (and, or, but, etc.) are clearly the product of the human mind and help us organize our thoughts and communicate, but they have no scientific basis. In an ideal world, our language would accurately reflect the real scientific distinctions in the world, which brings us back to Plato and the distinction between words and Forms or Ideas. According to Plato, the Demiurge used the Forms or Ideas to create the scientifically distinct things and species in the world, like a 3D printer. It is our job to adjust our language to reflect this reality and to recognize that definitions are independent of the human mind. However, no one in his right mind would equate a word with the Form of Idea. If a word is an arbitrary sound that represents a thing or species in the world, the Form or Idea is the algorithm or blueprint that makes the thing or species possible. Language, words, and alphabets are human creations, a product of mind. Thus, whether we agree with Plato or Aristotle, the one thing that both sides should agree on is that words do not exist in the Platonic realm.

The same idea applies to numbers. As we have seen, quantitative or mathematical thinking is possible without formal numbering systems. Just as our minds naturally conceptualize, our minds naturally quantify. A mother can distribute candy to her children without understanding numbers just as a tribe in the desert can load a sufficient amount of water onto camels to cross the desert without understanding numbers.

Just as in the case of language, the numbering system we use is as much a reflection of the human mind as it is a reflection of reality. The primary different between language and mathematics is that number is not predicated of anything in the world. There is a plurality of objects in the world and the widest abstraction for this idea is the category of quantity, but base-10, place-value numbers ultimately reflect the human mind and serve the purpose of bringing the vast quantity of objects in the world under the grasp of the human mind. And whereas it makes some sense to talk about the Demiurge using some type of Form or Idea of "tiger" in the Platonic realm to create tigers in the world, it makes no sense to say the Demiurge uses the Form or Idea of "2" to create pairs of objects in the world. Therefore, as was the case with letters and words, there are no numbers in the Platonic realm. The Platonic realm is theoretically capable of producing a plurality of objects in our world, like changing the copies from 1 to 2 on the 3D printer, but there are no numbers in the Forms of Ideas.

The key point is there is no contradiction between understanding mathematics in the context of potential infinity and solving problems on a continuum with real numbers. Infinity does not have to pre-exist as a complete set to solve problems. Numbers exist in our minds when we think them, and there is no limit to how many we can produce or how much we can divide them. Just as we make language as refined as we need for particular situations, without presuming the existence of an infinite set of words in the Platonic realm, we can solve any problem in mathematics with the level of precision we need without presuming the pre-existence of numbers we have never used before.

A potential or uninstantiated number does not have the same kind of existence as a potential or uninstantiated color because colors represent real frequencies in the world. A color frequency exists independent of the human mind, but a number does not. Thus, we see that Cantor's idea of infinity has two problems. First, as I have shown in this book, the idea of infinity (dimensionless points on a finite continuum) violates the law of contradiction. Second, all numbering systems are the product of the human mind with no independent existence in this world or a Platonic realm. This is why Plato said mathematical objects (numbers and shapes) had an intermediary existence between perception and the Forms or Ideas. I think what Plato meant to say was that numbers exist in our minds.

Thus, when we talk about the uncanny correlation between mathematics and the world, we should consider it from two perspectives: first, the regularity or patterns we perceive in the world; and second, the ability of numbers to accurately mirror or predict the world. We will address the latter point here. If we consider that number can be predicated of anything but is not predicable of anything, it should be clear why mathematics is so powerful. For example, if we combine two groups of 4 objects, the result will always be 8 objects. Unlike words, which apply to a limited part of the world, numbers apply to all things at all times. If a desert tribe of 100 needs one container of water per day per person for a 3-day trip across the desert, we can calculate with precision that the tribe will need 300 containers of water to cross the desert. Likewise, if we know that a strain of bacteria doubles in size every day, then if the population is 1 on day 1, then the population will be 64 on day 7. This does not explain why the population doubles every day, but it allows us to calculate the future population with precision. The same thing applies to more complex examples, like planetary orbits and acceleration. For example, we can use mathematics to accurately track the motion of a planet along an elliptical orbit or a falling object without understanding what makes the force of gravity possible. The quantitative aspect of reality is not the cause of what happens in the world, but we can use the science of quantity (mathematics) to predict what will happen in the future.

Non-discursive Thought

If numbers, letters, and words do not exist in a Platonic realm, then what can we say about what does exist in a Platonic realm (a non-material realm, the mind of God, etc.?) Of course, if you reject the possibility of a Platonic realm, then you can reject actual infinity and call it a day. You can sign up for the potential infinity club and enjoy the benefits. Before we answer that question, however, we should consider the relationship between numbers, letters, and words and the human mind because, at a minimum, we know that they exist in the human mind.

Suppose you were watching television one day and saw a journalist interview a famous poet, which included a question about the writing process. Imagine how shocked you would be if the poet said the creative process involved taking a list of random words from the dictionary and arranging them until he is satisfied with the result. After all, poems consist of words, which means the raw material of poems must therefore be words. Or suppose you were watching an interview of a scientist who discovered a formula to explain the universe. Imagine how shocked you would be if the scientist said the secret to his success is making a list of numbers, variables, and operations and arrange them until he is satisfied with the result. After all, formulas consist of numbers, variables, and operations. In both cases, you would probably reject the explanation as farcical. We have all experienced writing a letter or email to someone and thinking about what to write before we start writing, and then editing the text after to refine "the message." There is no doubt that our final product consists of words, but the individual words were not the raw material in the writing process. A lot of our thinking is not done with words.

To clarify the issue, what thought process do we use to make words and numbers? We do not initially use numbers to think about how to make a numbering system and we do not use numbers to think about how to solve a mathematical problem. Likewise, we did not initially use words to think about how to make words and we do not use words to think about the meaning of a text. Granted, our use of numbers and words over time will shape the way we think, and some unfortunate people might find themselves thinking exclusively in words and numbers, but the transition should never be complete, especially if we hope the maintain a capacity for art and creative thought.

However, when we transition from raw thought to meaningful sentences, one word at a time, we make the transition to discursive thought. For example, suppose we realize one day that we love someone and want to express our love. There is no doubt that we are thinking about the person we love, but there is also no doubt that our thought process does not involve thinking individual words in a discursive way. We might eventually say to the person, "You make my love bloom like a rose in spring!" but this sequence of 10 words is not identical with the feeling of love we have for the person. If someone else looked at the same person and said those same 10 words, the person probably would not suddenly fall in love with the person. The ten words are a reflection of something deeper and more profound, and there is virtually no limit to how this feeling can be expressed in words, evidenced by the steady flow of love songs and romantic comedies.

The same idea applies to mathematics. If an engineer thinks about how to solve a problem, he will jot notes and eventually come up with one or more formulas to solve the problem, often in such a way that the final formula was not part of the original thought process. I do not claim to fully understand this process, but what we can say with certainty is that a lot of our productive thinking is non-discursive.

If we think about what makes discursive though possible, we could use a word like mind. The human mind is capable of creating words and numbers, but mind is not reducible to words and numbers. However, this is the case with computers for which the algorithms and the "thinking process" of the computer are one and the same. This, in my opinion, is why artificial intelligence (AI) will always fall short of the human mind and will always be limited in terms of creativity.

Computers lack a capacity for non-discursive thought because discursive thought can never be reduced to 1s and 0s. For example, if we write a computer algorithm to drive a car, the activity of the car will be identical with the algorithm. The computer does not have a non-discursive meta-processing mind that monitors the processes of the computer and makes adjustments, unless this monitoring function is also an algorithm (written by a human mind). Just as important, the algorithm will simulate thought, but only because a human mind wrote the algorithm. In theory, an algorithm could be written to modify itself in response to new information, such as heat, but only because a human mind wrote the algorithm. Thus, when a computer "thinks," it is identical to its own discursive algorithms, which were written by a human mind. For the same reason that set theory will forever be incomplete, a computer will never truly think.

In the case of humans, there can be a great disparity between discursive language and the activity of the mind, both in terms of simplicity and complexity. A computer could never duplicate the scientific or mathematical successes of someone like Newton because these successes rely on intuition and non-discursive thought, which is something that a mind possesses but a computer does not. A computer can crunch the numbers, but cannot transcend its own programming. On the other hand, a computer could be programmed to write words with character and plot, which is not as impressive as it sounds. The true meaning of a novel transcends the individual words, characters, and plot (stories often have a grand metaphor to express meaning), and a computer narrative would be about nothing more than the individual words, characters, and plot. A computer could be programed to use specific metaphors, but the computer could never generate new metaphors or consider the entire narrative as a grand metaphor in a way that goes beyond the programming of a human mind.

I raise the idea of non-discursive thought with some hesitation because some people might take it the wrong way, in particular, the idea that we need not burden ourselves with the rigors of discursive thought if we can take comfort in the realm of non-discursive thought, mystical connotations and all. If non-discursive thought is the realm of creativity and intuition, then the more time we spend there the better. This is not true—in fact, the opposite is true. The negative kind of mystical thought is what happens when a mind is devoid of the rigors of discursive thought. This is not to say that all mystical thought is wrong, but it does mean that a proper refinement of our discursive thinking will improve our capacity for non-discursive thinking, and vice versa.

The brain is a like a muscle that has to be trained and used to develop its potential, and a good portion of this potential is designed to process discursive thinking. Consider a novel. A would-be writer might have a grand metaphor for a story, but the author will not be able to deliver if he has not trained and honed his discursive writing skills for many years. As Schopenhauer noted, novels use discursive language to stimulate a non-discursive, perceptual experiences in the mind of the reader. "For in plastic and pictorial art allegory leads away from what is given in perception, from the real object of all art, to abstract thoughts; but in poetry the relation is reversed. Here the concept is what is directly given in words, and the first aim is to lead from this to the perceptive, the depiction of which must be undertaken by the imagination of the hearer."[24] At the same time, I would advise against an unhealthy obsession with discursive thought that attempts to squeeze out all remnants of non-discursive thought.

Pattern Recognition

If the claim is that actual infinity is real because it exists in the human mind, then we can say it is true only in a trivial sense of the word. No one would disagree that the human mind can think about infinity, which is not the same as having an infinite thought, but this view misses the point because the idea of infinity has meaning only if it is true independent of the human mind and shapes the world we live in. When Cantor reflected on infinity and drafted his proofs, he was clearly thinking about something that transcended his own mind.

I am not aware of anyone who believes numbers are material objects that shape the world we live in, so I will not waste any time considering this option. Any discussion about the metaphysics of infinity necessarily takes us

[24] Schopenhauer, Arthur, *The World as Will and Representation*, 240.

to an immaterial realm, such as a divine mind or a Platonic realm. We can go all the way back to the pre-Socratics and the philosophy of Anaxagoras for the idea that the principle of all things is Nous or Mind. For atheists or materialists, there is no room for this immaterial realm and therefore no room for the metaphysics of infinity. According to atheism or materialism, the material world is the result of random chance and consists of material objects colliding in space. The human mind is an epiphenomenal accident, and the numbering system and mathematics is nothing more than a convenient human creation that just happens to correspond to the world.

Setting aside atheism and materialism for now, when we consider the metaphysics of infinity, we can consider complexity and the deep patterns of the universe. In the previous section we addressed the complexity of things that can be traced back to the work of a human mind, such as art and inventions. No one would doubt that many things are the work of a human mind, not random chance, but the issue is more controversial with nature, to include the humans who are capable of creating art and inventions. When we normally consider complexity and patterns, the general assumption is that like creates like and cannot transcend its own limitations.

For example, if we were to find a novel on a deserted island, we would be safe to assume that a human or other intelligent life wrote the book. Not only that, we could discern something about the level of intelligence based on the complexity of the text. If our child in first grade was asked to write a story and handed us a final product that read like Shakespeare, we would know that something was amiss because a human mind cannot create something that transcends its own limitations, although purposeful writing and thinking over time can push us beyond our current limitations to fulfill more of our potential.

However, when we make the transition from humans to the universe, many people, especially atheists and materialists, would argue that the same rules do not apply. We are told that evolution (random genetic mutations) is sufficient to account for the complexity and deep patterns we see in the world, to include the humans who possess a capacity to write novels. A mind is required to write a novel, no question about it, but a mind is apparently not required to make humans who are capable of writing novels. The secret to this mystery we are told is time, lots of time, billions of years, even though it is easy to use mathematics to prove that certain examples of complexity we see in the world cannot be achieved via random processes.

The purpose of this section is not to assess whether or not evolution is capable of producing the complexity we see in the world, but rather to consider it in the context of infinity. The question is, in a Platonic realm, would an infinite set of real numbers be required to produce the complexity we see in the world? The simple answer is no, because living organisms grow in complexity via edits to DNA, our code or algorithm, not via numbers, just as a short story can be transformed into a novel with a series of purposeful edits. At no step along the way are numbers involved. Whether a human mind edits a novel or the DNA code gets edited from a Platonic realm, no numbers are required to achieve new levels of complexity.

In the case of deep patterns, the situation is different because the search for deep patterns is often the search for the simplicity that makes complexity possible. For example, Newton studied the complexity of the universe to see the deep pattern of gravity and developed a simple yet powerful formula.

$F = G(m_1 x m_2)/r^2$

What this means is that the force of gravity (F) between any two objects is the mass of the first object (m_1) times the mass of the second object (m_2) divided by the distance between them squared (r^2), multiplied by a gravitational constant (G). This is one of many formulas that are capable of accurately describing how objects behave in the universe. From physics to thermodynamics to quantum theory to string theory, we have simple formulas that bring the vast complexity of the universe within our grasp. The idea that all of these formulas for deep patterns could be the result of random chance seems unlikely, but even if we presume a Platonic realm, it raises questions about whether an infinite set of real numbers would be required to make these formulas. In this case, the answer is easier than the case of complexity because formulas do not require numbers. We can plug numbers into formulas to solve problems, but the numbers are not required to make formulas or blueprints.

If we entertain the possibility of a Platonic realm, then we should ask how we account for the complexity and deep patterns of the world and what role, if any, there is for infinity. Beginning with infinity, one way to think about it is what happens when the infinity of division achieves its goal and we finally reach the singularity of dimensionless ranges? The truth is we never do: Aristotle was right about potential infinity. Just as Cantor said a discrete counting process never reaches infinity, a discrete divisibility process never reaches infinity, and an infinite set of dimensionless points lack extension (0 + 0 + 0 + ... + 0 = 0) and will never fill the space of a finite continuum. The only way to cross the abyss from the finite to the infinite is by decree, the Axiom of Infinity, but if we make this leap, we find pure mind or thought devoid of numbers or distinctions of any kind. Just as the human mind does not rely on numbers and words to think about numbers and words, it would not make sense for the Platonic realm to rely on numbers and words to create the world.

Conclusion

I am mindful that my analysis and conclusions probably will not be welcomed by the Cantor camp and that the potential infinity camp, of which I am a member, will probably nod approvingly but avoid saying too much for fear of being labeled a "crank." Like the current political divide, most of the debate boils down to the fact that both sides are speaking two different languages and basing their ideas on two different sets of assumptions. In this case, most of the debate boils down to whether the Axiom of Infinity is valid. Admittedly, there were moments while writing this book when I paused to consider the possibility that I was wrong, but the fact that the Axiom of Infinity involves a logical contradiction kept me moving true north. This is no place to rehash the arguments of this book, but if my analysis has helped anyone from the Cantor camp grasp how fragile their position is, or to even consider the possibility that the potential infinity camp position wins the day, then my effort will have been worthwhile.

Appendix: Simplified Proofs

1. Proof that set theory does not provide a foundation for mathematics:

If set theory provides a foundation for mathematics, then set theory would generate every known mathematical truth (complete) and would never generate contradictions (consistent). First, according to Gödel's incompleteness theorem, set theory is incomplete. There are known mathematical truths that set theory cannot generate, and it is impossible to add more axioms to set theory to resolve this problem. Second, set theory generates contradictions, which means set theory is inconsistent. For example, Cantor's claim that $\aleph < 2^{\aleph}$ generates a contradiction (see below), and the idea of actual infinity generates a contradiction (see below). Therefore, set theory does not provide a foundation for mathematics.

QED

2. Proof that Cantor's one-to-one correspondence or isomorph methodology is invalid for determining the cardinality of a set:

According to Cantor, if two infinite sets can be put into a one-to-one correspondence or isomorph, and one set is a proper subset of the other set, they are the same size or cardinality. For example, if for every natural number we can list an even number (the even numbers are a proper subset of the natural numbers), and this process continues ad infinitum, then the two sets are the same size or cardinality. In other words, Cantor claims the fact that we can start the process of pairing up an infinite set with a proper subset of itself means the two sets are the same size or cardinality. However, the natural numbers are a proper subset of the real numbers, we can start the process of pairing up the two sets of numbers, and the process continues ad infinitum, but according to Cantor, the set of real numbers is larger than the set of natural numbers. Thus, the one-to-one correspondence or isomorph methodology is invalid for determining the size or cardinality of a set.

QED

3. Proof that the idea of actual infinity is incompatible with a finite continuum.

According to set theory, every finite continuum contains an infinite set of points with the cardinality of the set of real numbers ($\aleph_1$). On the one hand, if the points are dimensionless, then the set of infinite points lacks extension (0 + 0 + 0 ... + 0 = 0) along a finite continuum. On the other hand, if the points have dimension, no matter how small, then the infinite set of points will necessarily extend to infinity, making a finite continuum impossible. There is no way to evenly distribute an infinite set of points along a finite continuum. Therefore, the idea of an infinite set of points is incompatible with a finite continuum and violates the law of contradiction by claiming that points both have and lack dimension at the same time.

QED

4. Proof that the rational numbers and the real numbers have the same cardinality.

First, let us construct the set of rational numbers a decimals one place-value decimal number at a time.

1. 0.0, 0.1, 0.2, 0.3 ... 0.9.

2. 0.00, 0.01, 0.02, 0.03 ... 0.99.

3. 0.000, 0.001, 0.002, 0.003 ... 0.999.

...

After the first decimal place we have 10 rational numbers; after the second decimal place we have 100 rational numbers; after the third decimal place we have 1,000 rational numbers, and so on. For every additional place-value number in the decimal, the size or cardinality of the set grows by a factor of 10. To generalize, given that each place-value number in the decimal represents one of 10 possible numbers (0–9), then the size or cardinality of rational numbers at any given point in the process is 10^n, where 10^n represents the size or cardinality of the set of rational numbers with n decimal places. According to Cantor, the cardinality of the rational numbers is $\aleph$.

Second, let us construct the set of real numbers by considering the set of infinite decimals. Given that the place-value numbers in the infinite decimals are natural numbers, the size or cardinality of the set of place-value numbers in each infinite decimal is $\aleph$. And, given that each place value number in the infinite decimal represents one of 10 possible numbers (0 –9), the cardinality of the set of real numbers is $10^{\aleph}$. (Cantor used the power set of the natural numbers and infinite binary decimals to claim the cardinality is $2^{\aleph}$.) Thus, given that the cardinality of the real numbers ($10^{\aleph}$) is larger than the cardinality of the rational numbers ($\aleph$), Cantor claims the real numbers are not countable.

However, given that the place-value numbers in the infinite decimal are natural numbers and have a cardinality of $\aleph$, then in the case of the rational numbers, as n goes to infinity, 10^n goes to $10^{\aleph}$, which is the same cardinality as the real numbers. Thus, the rational numbers as decimals and the real numbers have the same cardinality if actual infinity is real (the Axiom of Infinity). Additionally, because the cardinality of the rational numbers as fractions ($\aleph$) and the rational numbers as decimals ($10^{\aleph}$) are different, we prove that set theory is inconsistent.

QED

5. Proof that $\aleph \times \aleph \neq \aleph$ and $\aleph + \aleph \neq \aleph$.

According to Cantor, $\aleph \times \aleph = \aleph$ and $\aleph < 2^{\aleph}$. In other words, not all infinities are the same size and the arithmetic operations do not apply to infinite sets. If we assume this is true, then the following are also true (expressing exponents as multiplication).

$\aleph \times \aleph \times \aleph \times \ldots \times \aleph = \aleph$

$2 \times 2 \times 2 \times \ldots \times 2 = 2^{\aleph}$ (per Cantor's use of infinite binary decimals).

If we recall Cantor's claim that $\aleph < 2^{\aleph}$, then the following is true.

$\aleph \times \aleph \times \aleph \times \ldots \times \aleph < 2 \times 2 \times 2 \times \ldots 2$

Given that these two sets have the same cardinality ($\aleph$), we can put them into a one-to-one correspondence or isomorph and simplify.

$\aleph \cancel{\times \aleph \times \aleph \times \ldots \times \aleph} < 2 \cancel{\times 2 \times 2 \times \ldots 2}$

This in turn means the following is true.

$\aleph < 2$

Therefore, only one of the following can be true: $\aleph < 2$, $\aleph \times \aleph \times \aleph \times \ldots \times \aleph \neq \aleph$, or $2 \times 2 \times 2 \times \ldots \times 2 \neq 2^{\aleph}$. As the former and the latter are clearly false, we can conclude that only the middle term is true, which means $\aleph \times \aleph = \aleph^2$, $\aleph \times \aleph \times \aleph = \aleph^3, \ldots$, and $\aleph \times \aleph \times \aleph \times \ldots \times \aleph = \aleph^{\aleph}$. Therefore, because exponents are a form of multiplication and exponents apply to infinite sets (according to Cantor), then multiplication also applies to infinite sets.

Further, given that multiplication is a form of addition (for example, $3^2 = 3 \times 3 = 3 + 3 + 3$ or $5^2 = 5 \times 5 = 5 + 5 + 5 + 5 + 5$) the following is true.

$\aleph + \aleph + \aleph + \ldots + \aleph = \aleph^2$

This in turn means the following must be true, which contradicts Cantor.

$\aleph + \aleph = 2\aleph$, $\aleph \times \aleph + \aleph = 3\aleph, \ldots, \aleph + \aleph + \aleph + \ldots + \aleph = \aleph\aleph = \aleph^2$

If multiplication applies to infinite sets, then division applies because division is the inverse of multiplication. This means $\aleph/2 < \aleph$, which means the set of even numbers is half the size as the set of natural numbers.

In short, by claiming that the cardinality of the real numbers is larger than the cardinality of the natural numbers ($\aleph < 2^{\aleph}$), Cantor discredits his own arithmetic of infinity. The only way for Cantor to salvage his arithmetic of infinity is to reject this inequality, in which case all infinite sets are the same size.

QED

6. An irrational number sighting:

For any circle, the ratio of the circumference to the diameter is π. If we imagine a circle with a circumference of 1, then the diameter is $1/\pi$, such that $1 \div 1/\pi = \pi$. Now, if we take the circumference, bend it into a straight line (gray), and impose the diameter onto it (black), we have the following:

The dimensionless point separating the black and gray ranges represents $1/\pi$, an irrational number. There it is. We can see it. Thus, it makes sense to say a point representing an irrational number exists along a continuum, which represents the ratio of the circumference to the diameter, but only after we arbitrarily defined the circumference as 1. Of note, this dimensionless point did not exist until we imposed the black line onto the gray line, there is no way to express this geometric ratio as a rational number, and there is no limit to how many other dimensionless points we can generate by changing the length of the black range, ad infinitum.

Selected Bibliography

Alexander, Amir, *Infinitesimal: How a Dangerous Mathematical Theory Shaped the Modern World* (New York: Scientific American/Farrar, Straus and Giroux, 2014).

Aristotle, *The Basic Works of Aristotle*, edited by Richard McKeon (New York: Random House, 1941).

Bagaria, Joan, "Set Theory," *The Stanford Encyclopedia of Philosophy* (Winter 2014 Edition), Edward N. Zalta (ed.), URL = ‹http://plato.stanford.edu/archives/win2014/entries/set-theory/›.

Balaguer, Mark, "Fictionalism in the Philosophy of Mathematics," *The Stanford Encyclopedia of Philosophy* (Summer 2015 Edition), Edward N. Zalta (ed.), URL = ‹http://plato.stanford.edu/archives/sum2015/entries/fictionalism-mathematics/›.

Bell, John L., "Continuity and Infinitesimals," *The Stanford Encyclopedia of Philosophy* (Winter 2014 Edition), Edward N. Zalta (ed.), URL = ‹http://plato.stanford.edu/archives/win2014/entries/continuity/›.

Bridges, Douglas and Palmgren, Erik, "Constructive Mathematics," *The Stanford Encyclopedia of Philosophy* (Winter 2013 Edition), Edward N. Zalta (ed.), URL = ‹http://plato.stanford.edu/archives/win2013/entries/mathematics-constructive/›.

Bueno, Otávio, "Nominalism in the Philosophy of Mathematics," *The Stanford Encyclopedia of Philosophy* (Spring 2014 Edition), Edward N. Zalta (ed.), URL = ‹http://plato.stanford.edu/archives/spr2014/entries/nominalism-mathematics/›.

Colyvan, Mark, "Indispensability Arguments in the Philosophy of Mathematics," *The Stanford Encyclopedia of Philosophy* (Spring 2015 Edition), Edward N. Zalta (ed.), URL = ‹http://plato.stanford.edu/archives/spr2015/entries/mathphil-indis/›.

Copleston, Frederick, *A History of Philosophy Volume I: Greece and Rome* (New York: Image Books, 1993).

Dauben, Joseph Warren, "Georg Cantor: His Mathematics and Philosophy of the Infinite" (Cambridge: Harvard University Press, 1979).

Franklin, James, *An Aristotelian Realist Philosophy of Mathematics: Mathematics as the Science of Quantity and Structure* (New York: Palgrave Macmillin, 2014).

George, Alexander; Velleman, Daniel J., *Philosophies of Mathematics* (Malden, MA: Blackwell, 2002).

Havil, Julian, *The Irrationals: A Story of the Numbers You Can't Count On* (Princeton: Princeton University Press, 2012).

Hume, David, *A Treatise of Human Nature* (New York: Barnes & Noble, 2005).

Iemhoff, Rosalie, "Intuitionism in the Philosophy of Mathematics," *The Stanford Encyclopedia of Philosophy* (Spring 2015 Edition), Edward N. Zalta (ed.), URL = ‹http://plato.stanford.edu/archives/spr2015/entries/intuitionism/›.

Linnebo, Øystein, "Platonism in the Philosophy of Mathematics," *The Stanford Encyclopedia of Philosophy* (Winter 2013 Edition), Edward N. Zalta (ed.), URL = ‹http://plato.stanford.edu/archives/win2013/entries/platonism-mathematics/›.

Mancosu, Paolo, "Explanation in Mathematics," *The Stanford Encyclopedia of Philosophy* (Summer 2015 Edition), Edward N. Zalta (ed.), URL = ‹http://plato.stanford.edu/archives/sum2015/entries/mathematics-explanation/›.

Mendell, Henry, "Aristotle and Mathematics," *The Stanford Encyclopedia of Philosophy* (Winter 2008 Edition), Edward N. Zalta (ed.), URL = ‹http://plato.stanford.edu/archives/win2008/entries/aristotle-mathematics/›.

Merzbach, Uta C.; Boyer, Carl B., *A History of Mathematics*, third edition (Hoboken, NJ: John Wiley & Sons, 2011).

Núñez, Rafael E., "Creating mathematical infinities: Metaphor, blending, and the beauty of transfinite cardinals" (*Journal of Pragmatics*, 37 (2005), 1717 - 1741).

Paseau, Alexander, "Naturalism in the Philosophy of Mathematics," *The Stanford Encyclopedia of Philosophy* (Summer 2013 Edition), Edward N. Zalta (ed.), URL = ‹http://plato.stanford.edu/archives/sum2013/entries/naturalism-mathematics/›.

Plato, *Complete Works*, edited by John M. Cooper (Indianapolis: Hackett Publishing, 1997).

Reichenbach, Bruce, "Cosmological Argument," *The Stanford Encyclopedia of Philosophy* (Winter 2016 Edition), Edward N. Zalta (ed.), URL = <https://plato.stanford.edu/archives/win2016/entries/cosmological-argument/>.

Rodych, Victor, "Wittgenstein's Philosophy of Mathematics," *The Stanford Encyclopedia of Philosophy* (Summer 2011 Edition), Edward N. Zalta (ed.), URL = <http://plato.stanford.edu/archives/sum2011/entries/wittgenstein-mathematics/>.

Ross, Sir David, *Aristotle*, sixth edition (New York: Routledge, 1995).

Russell, Bertrand, *The History of Western Philosophy* (New York: Simon & Schuster, 1972).

—*Introduction to Mathematical Philosophy* (Digireads.com Publishing, 2010).

Schopenhauer, Arthur, *On the Fourfold Root of the Principle of Sufficient Reason* (La Salle, IL: Open Court, 1999).

_____, *The World as Will and Representation* (New York: Dover Publications, 1969).

Shabel, Lisa, "Kant's Philosophy of Mathematics," *The Stanford Encyclopedia of Philosophy* (Fall 2014 Edition), Edward N. Zalta (ed.), URL = <http://plato.stanford.edu/archives/fall2014/entries/kant-mathematics/>.

Shapiro, Steward, *Thinking About Mathematics: The Philosophy of Mathematics* (New York: Oxford University Press, 2011).

Zenkin, Alexander A., "Logic of Actual Infinity and G. Cantor's Diagonal Proof of the Uncountability of the Continuum," *The Review of Modern Logic* (Volume 9, Numbers 3 & 4, December 2003 - August 2004, Issue 30).

Index

O

P

R

S

T

U

W

Z

Printed in the United States
By Bookmasters